SECRETS OF A GOOD **DIGESTION**

*This volume is one of a series designed to familiarize readers
with the latest advances in medical science as a guide in
maintaining their own health and fitness.*

SECRETS OF A GOOD **DIGESTION**

By Oliver E. Allen

AND THE EDITORS OF TIME-LIFE BOOKS

LIBRARY OF HEALTH / TIME-LIFE BOOKS / ALEXANDRIA, VIRGINIA

THE AUTHOR:
Oliver E. Allen, whose previous books cover a range of subjects from wild flowers to windjammers, is the author of the volume *Building Sound Bones and Muscles* in the Library of Health. He was formerly an editor on the staff of Time-Life Books.

THE CONSULTANT:
Dr. Nathaniel Cohen is Chief of the Gastrointestinal Clinic at New York's Mount Sinai Hospital, and Associate Clinical Professor of Medicine at Mount Sinai School of Medicine. A graduate of Harvard Medical School, Dr. Cohen is also Attending Gastroenterologist at the Jewish Home and Hospital for the Aged and at the Bronx Veterans Administration Medical Center in New York, and has written extensively on gastrointestinal disorders.

For information about any Time-Life book, please write:
Reader Information, Time-Life Books,
541 North Fairbanks Court, Chicago, Illinois 60611.

First printing. Printed in U.S.A.
Published simultaneously in Canada.
School and library distribution by Silver Burdett Company, Morristown, New Jersey.

TIME-LIFE is a trademark of Time Incorporated U.S.A.

Library of Congress Cataloguing in Publication Data
Allen, Oliver E., 1922-
　　Secrets of a good digestion
　　(Library of Health)
　　Bibliography: p.
　　Includes index.
　　1. Gastrointestinal system—Diseases—Prevention.
　　2. Gastroenterology—Popular works.
　　I. Time-Life Books.　II. Title.　III. Series.
RC802.A37　　616.3　　82-694
ISBN 0-8094-3792-9　　AACR2
ISBN 0-8094-3791-0 (lib. bdg.)
ISBN 0-8094-3790-2 (retail ed.)

CONTENTS

Off the table and into the body

Imagine the discomfort of poor Mr. Polly. The rotund hero of a 1909 comic romance by the English novelist H. G. Wells, he was a compulsive trencherman. At one memorable sitting, Mr. Polly managed to work through a heaping platter of cold pork, a dish of cold potatoes mixed with pickles, onions, cauliflower and capers, a cold suet pudding with treacle, along with hard cheese and several slabs of leaden, grayish bread—the whole washed down with a mighty jugful of beer. This huge repast, said Wells, caused "wonderful things" to occur in Mr. Polly's body. "It must have been like a badly managed industrial city during a period of depression," Wells wrote, "agitators, acts of violence, strikes, the forces of law and order doing their best, rushings to and fro, upheavals, the Marseillaise, tumbrils, the rumble and the thunder of the tumbrils."

Mr. Polly survived the tumult, and so do most people who indulge themselves, however extravagant the menu. The ability of human beings to cope with the most outrageous demands of appetite is wonderful indeed. Homo sapiens is omnivorous. Unlike the cow munching its cud, or the hummingbird sipping nectar, people can eat almost anything, and commonly do—grains, roots, leaves, berries, fish, meat and milk, all containing complex mixtures of potential nutrients. Only such woody fare as grass stems and tree bark are truly inedible: The human system lacks the chemicals needed to process cellulose. All other foodstuffs—the gallimaufry of starches, sugars, fats and proteins that makes up diet—are regularly consumed and converted into energy and tissue within the body. The first stage in this conversion is digestion. Each mouthful of food must be pulverized, dissolved or emulsified, and broken down chemically into submicroscopic units that can be taken into the bloodstream.

As food makes its way down the gullet, or esophagus, and through the 30-foot length of the alimentary canal, it is churned and bombarded with a well-timed sequence of agents that change it into one form after another—usually for good purpose, sometimes for ill. Glands in the mouth pour in saliva; cells in the linings of the stomach and intestines add potent digestive juices. Two of the body's largest organs, the liver and pancreas, add other essential secretions. Then, when the nutrients have been extracted and absorbed into the blood for distribution throughout the body, the residue is expelled. The process goes on, day in and day out, year after year, for a lifetime, with very little conscious effort.

For all its complexity, it is astonishing how smoothly the process goes—usually. But every now and again something goes awry, and the results are never pleasant. Under the impact of foods that are too rich or exotic, or of meals taken too fast or too gluttonously, the system rebels. Any number of viruses or bacteria—some ordinarily friendly—can attack the gut to bring sudden misery. A quarrel or a bad day at work can cause upset: No system of the body is so immediately sensitive to emotional stress as the digestive tract.

The results may be as minor and brief as a mild stomachache, generally curable by a simplified diet. But they can also be brutally severe, with devastating bouts of nausea and

In this 19th Century Japanese woodcut, tiny figures illustrate an old Oriental theory of digestion: Food moved from the stomach (above the brownish intestines), to the mixing vat of the liver, the fiery caldron of the spleen, and the heart, "Lord of Entrails."

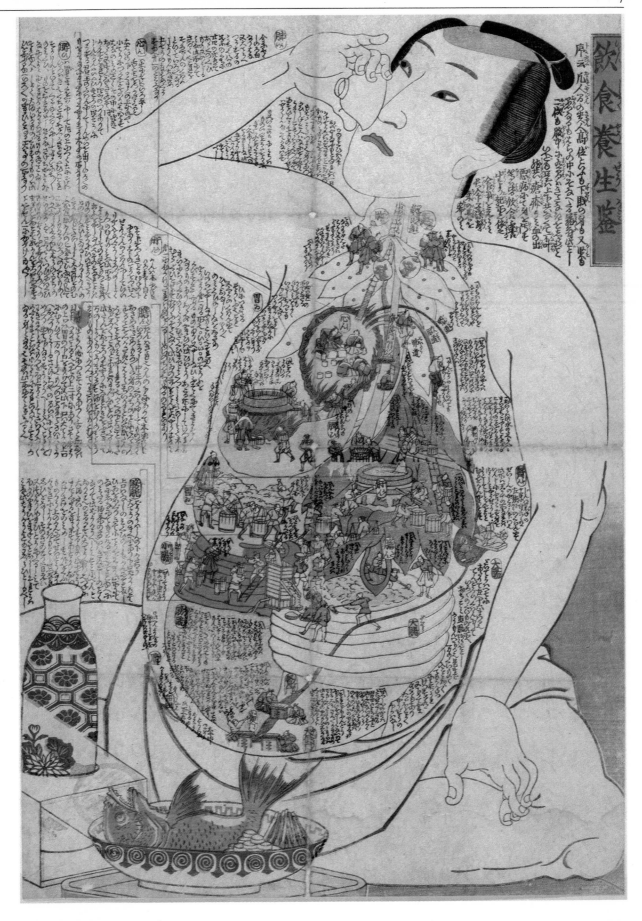

vomiting, acute abdominal pain and diarrhea. Any one of these symptoms can send the victim to bed and, depending on the circumstances, can indicate anything from food poisoning to a disease more dangerous. The sharp, stabbing pain of an appendicitis attack, or of a hitherto-undetected bleeding ulcer will send the sufferer posthaste to the hospital. So too may the nausea and fever of a severely inflamed liver, or the ravages of various painful and unpleasant afflictions of the lower bowel, such as diverticulitis and ulcerative colitis.

Often there is no discernible cause, no organic disruption that shows up in the usual routine of blood tests, X-rays, probings and inspections. Doctors call such ailments "functional," meaning that the patient's digestive apparatus simply does not work as it should and they cannot figure out why. Food moves through too slowly, resulting in constipation, or too quickly, yielding chronic diarrhea. Or the gut is puffed up by quantities of internal gas, making for flatulence and bloating. Or it just hurts somewhere along the way.

The digestive system's central role in well-being and the mystifying frequency with which it slips out of kilter give its ills major importance. Serious stomach and intestinal disorders, such as ulcers and lower-bowel diseases, account for 10 per cent of the days lost through illness by American adults, and 15 per cent of all admissions to general hospitals. The economic burden is astonishing. Direct medical expenses exceed $20 billion in any given year, and the cost in lost wages and related effects runs over $35 billion. Only heart disease and accidents or violence take a greater toll.

No wonder that maintaining a good digestion concerns so many people. Most sufferers attempt to treat themselves, lavishing upwards of a billion dollars annually on a vast array of over-the-counter drugs. Many of these medicines are very useful, but others are not really needed and some can bring harm. Nearly all digestive ills pass quickly and completely. Most others are amenable to home treatment. Many can be prevented by changes in eating habits, or countered by a few prompt corrective steps. Others are now cured by new drugs, or are yielding to new therapies based on better understanding of the digestive process. New diagnostic tools let doctors isolate ailments more effectively. Today more than ever, there is a growing confidence that most digestive disorders can be dispatched swiftly through medical expertise, or successfully managed by the patient on his own.

When the digestion operates as it should—which is most of the time—it is a marvel of toughness and resiliency. Its defenses are formidable. Certain of its secretions are powerful enough to cause serious chemical burns; that they do not attack the system itself is seen to by other, protective agents that continually bathe the lining of the digestive tract. Thus human beings can extract the nourishment from animal intestines, such as sausage casings, without devouring their own intestinal walls.

At the same time that the system protects itself, it is also protecting its owner. The vomiting that may follow eating food that is over-rich, or has started to spoil, serves to guard the body against the possible toxins of decay. If ingested food fails inspection by the sensitive cells in the lining of the digestive tract, it can quickly be spat out, vomited up or—if it has already reached the large intestine—expelled at the far end. The rejects never enter the body proper to cause harm so long as they are in the digestive tract: They are not really inside the body. The tract, as it winds through the midsection, is still physiologically a part of the body's surface—a kind of interior skin between the body and whatever it takes in. Commented Dr. Benjamin Miller of the University of Cincinnati, "As long as the food is still in the digestive tract it might as well be on the table."

Getting food off the table, in Dr. Miller's sense, is the last step in digestion. To prepare food for that final step, digestive activity goes on automatically, with little conscious attention, and continuously day and night, in overlapping cycles: At dinnertime, for example, lunch and breakfast are both still undergoing treatment somewhere down the line. Normally, it takes at least 48 hours for a meal to make its way through the 30-foot length of the digestive tract.

Why your mouth waters

In a sense, digestion begins even before a bite of a meal is taken. The famous Russian physiologist Ivan Pavlov demonstrated in his experiments with dogs that the mere suggestion of food provokes a response that starts the mouth watering. The salivary glands along the jaw and underneath the tongue begin secreting a copious flow, which will moisten each mouthful for easy chewing, and will start the chemical breakdown of the food itself. At the same time, the stomach starts secreting other fluids that will continue the digestive process. Thus the system is already off and running in expectation of the first morsel.

As the saliva is secreted, it spreads out over the lining of the mouth. This same kind of surface lines the entire digestive tract. It is a thin membrane called the mucosa; it is pinkish because of the blood vessels immediately underneath, and wet because it exudes a film of mucus. The mucus is both a lubricant and a shield. It keeps the tract slippery to speed the passage of food, and it helps fend off attack from the digestive juices.

With the first bite, the teeth begin the process of grinding and tearing the food to pieces, starting a breaking-down operation that will be continued for some time in other parts of the tract. The front teeth can deliver a force as great as 55 pounds, the back molars up to 200 pounds, and they reduce the food to a fine-textured mush.

It is not necessary to chew each mouthful 32 times, as William Gladstone, Prime Minister during Queen Victoria's reign, believed, but thorough chewing does aid digestion. It breaks up the indigestible cellulose that covers the nutrients in raw fruits and vegetables. And even though the stomach juices can disintegrate the toughest chunk of steak, preliminary grinding speeds the process, and makes for a smooth, untroubled passage through the tract.

Moreover, chewing permits a good test for palatability. The 10,000 or so taste buds on the tongue not only promote enjoyment, but they also help you select foods that are good for you and avoid ones that are bad for you—they detect rot that could bring misery later on. This trial by taste is full of personal quirks—an unhappy childhood encounter with rhubarb, say, may cause this delicacy to be rejected ever after. Nor is flavor an infallible clue. The Amanita mushrooms with which Lucrezia Borgia is said to have poisoned her victims were, by all accounts, delicious. Nonetheless, the

Resembling a rose in a sea of petals, the object in the middle of this photomicrograph is actually a minute bump, called a fungiform papilla, on a rat's tongue; at its center is the tiny hole of a taste-bud pore. Human taste-bud pores, like the rat's, are clustered on the papillae. Nerve receptors inside each pore register one of the four basic tastes—sweet, sour, bitter and salty. The smaller projections, called filiform papillae, provide friction to move swallowed food backward in the throat.

taste buds do provide a first defense against gastric woe.

Chewing further encourages the flow of saliva. This vital fluid performs several functions, such as cleansing the mouth of the debris that decays teeth, and moistening the food particles. It also starts the process of chemical transformation, for it contains an enzyme, or biological catalyst, called amylase. Amylase acts on the complex molecules of starch, as in rice and potatoes, and breaks them down into simpler sugar compounds that can be absorbed by the body. The process can easily be detected: If you take some oyster crackers and chew them long enough, you will find that they presently turn sweet. Good saliva flow is essential to well-being; its cessation can lead to foul breath in an hour and tooth cavities in a week. For most people the threat is nonexistent, as the mouth normally produces more than a quart of saliva every day.

When a mouthful of food has been ground into mush, the tongue shapes it into a ball and thrusts it back toward the throat for swallowing. This is the last voluntary act in the digestive sequence. It is also more complex than it seems. The throat is a three-way funnel, with openings leading not only to the esophagus, where the mouthful of food is supposed to go, but also to the windpipe and nasal passages. If a person is to swallow effectively, the latter two openings must be closed, and normally a flap of tissue will move in to cover each. (The upward jiggle of the Adam's apple, a bit of cartilage that shields the top of the windpipe, is a sign that its flap, the epiglottis, is doing its job.)

On occasion the act of swallowing may misfire, however. Taking in a gulp of air at just the wrong moment may make the epiglottis stay open when it should be closed. A piece of food may then enter the windpipe, choking it. Coughing will usually bring up the offending article. When it does not, the cutoff of air can be dangerous—the victim may die of suffocation within minutes. (Because the crisis often occurs in a restaurant and its symptoms resemble those of a heart attack, it has been called the café coronary—though it does not involve the heart.) Swift first aid is vital, but the measures to be taken are in dispute. Some experts favor the old-fashioned slap on the back; others urge a sharp upward squeeze around the middle—the so-called Heimlich maneuver; and the

The lessons from Pavlov's dogs

The dogs Dr. Ivan Pavlov used in his experiments had humble origins—most were strays picked up by street thieves in St. Petersburg. But so great was their impact on digestive science that the Russian physiologist had a monument to The Unknown Dog erected outside his laboratory, dedicated to ''man's helper and friend from prehistoric times.'' Dr. Pavlov reciprocated the friendship, using humane procedures in all his experiments.

In his studies Dr. Pavlov collected gastric juices *(right)* and saliva *(bottom right)* through openings in the dogs' stomachs and salivary glands. For a time he sold leftover gastric juices as a digestive remedy—''incomparably superior,'' he boasted. But his concoction was not popular, partly because it was so acid it had to be sipped through a glass tube to prevent damage to teeth.

Dr. Ivan P. Pavlov appears above in a portrait dated 1904, the year he received the Nobel Prize for medicine and physiology. His early studies on the digestive processes of dogs foreshadowed the later work that brought him his greatest fame: Pavlov showed that dogs could be taught to salivate on signal even in the absence of food, the normal stimulus for salivation.

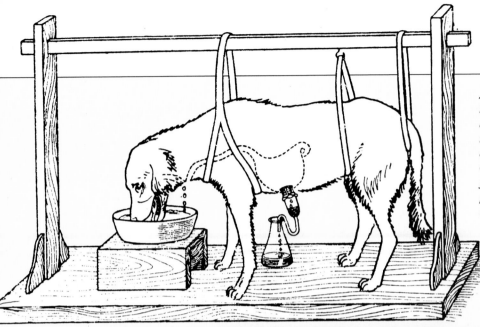

A diagram from one of Dr. Pavlov's books depicts an experiment in which the esophagus of a dog was surgically rerouted so that food, when swallowed, fell out of an opening in the dog's throat. Gastric juices nevertheless dripped into a jar, from an opening in the stomach, proving that digestive glands are controlled by the brain, not by the stomach.

A colleague distracts a restrained dog while Dr. Pavlov (foreground) observes the collection of saliva in a tube in the animal's cheek—a technique that was also employed in later psychological studies. Pavlov marveled at the salivary glands' ''high degree of adaptability,'' noting that different kinds and amounts of juices were secreted for different foods.

American Red Cross recommends trying both *(page 13)*.

Assuming no such mishaps, the food slips easily into the esophagus, a mucosa-lined oval tube some 10 inches long and about an inch wide that leads down to the stomach. In most people the tube is quite straight, a configuration of special use to sword swallowers *(below)*.

The act of swallowing triggers a succession of rhythmic muscle contractions in the esophagus itself. These contractions, known as peristalsis, push the food along, like toothpaste through a tube, at about an inch a second; they are a key factor throughout the digestive process. Peristalsis moves each mouthful through the system, the arrival of food at any

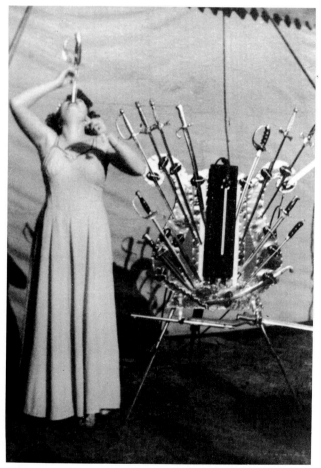

Sword swallower Leatha Smith demonstrates her magic in a 1940s circus sideshow. In actuality, sword swallowers do not swallow at all; the performer simply inches the instrument down the throat and into the esophagus. The trick is learning to suppress the gag reflex that generally constricts throat muscles when a large object enters the passageway.

section of the tract setting off spasms to shove it farther down the line. Without such contractions, there would be no way for substances to travel uphill, as they must at several spots in the abdomen. Peristalsis enables you to eat while standing on your head, if you wish, and it allows grazing animals to bring food from their lowered heads up to their stomachs.

As a rule peristalsis takes place spontaneously and with total efficiency. But cancer or injury can cause the muscles of the esophagus to weaken; then a bit of food will take a long time to descend or—if the morsel is a large one—may even become stuck. Farther along the digestive tract, peristalsis that is too slow results in constipation; when it is too fast it brings on diarrhea.

Food usually takes less than 10 seconds to be pushed by peristalsis to the foot of the esophagus, where it meets a particularly strong muscle ring, the lower esophageal sphincter. Guarding the entrance to the stomach, below the heart, this sphincter is one of several such portals in the digestive tract. Normally it stays closed until the arrival of food.

At the point where the esophageal sphincter stands guard, the esophagus passes through the diaphragm, a sturdy partition of muscle separating the chest cavity from the abdomen. In some people the opening, or hiatus, in the diaphragm becomes stretched—by obesity, pregnancy or tight clothing, among other things—and the top of the stomach pushes up through it. The result is a hiatal hernia. Generally the condition is symptomless and harmless. But when the sphincter is dislocated in this way it weakens, permitting the contents of the stomach to back up into the esophagus. The backflow of digestive juices will cause heartburn—so dubbed because it is near the heart. Although heartburn can arise in other ways—the sphincter may be weakened by many causes, such as smoking, caffeine and fatty foods—many people who blame their heartburn on simple indigestion actually have hiatal hernia.

What the stomach does

When the sphincter is working properly, it opens just before food reaches it, dropping the food into the stomach. Up to this time, a mouthful of food has not changed much from its mushy state except for a very preliminary breakdown of starches. It is now about to undergo considerable transformation. The stomach, situated slightly to the left side of the body and partly tucked behind the lower ribs, is an efficient device with a twofold mission. One is to serve as a reservoir, storing a meal's worth of food so that it can be admitted in steady, manageable amounts to the intestine. The other is to continue the process of digesting that was begun in the mouth, but on a much more intensive basis—by churning the food and mixing it with some extraordinarily powerful chemicals. These fluids, known collectively as the gastric juices, are secreted by the specialized cells of the mucosa lining the stomach.

The stomach, roughly J-shaped, is big, as much as 12 inches long and capable of holding a quart or more of food and drink. Much greater quantities are put into it by extremely heavy eaters. One such outstanding trencherman was Edward VII, England's monarch before World War I. The King, wrote John Stewart Collis, "began his day with a glass of milk in bed before coming down to breakfast which consisted of platefuls of bacon and eggs, haddock and chicken, toast and marmalade, with coffee. After one hour out shooting he returned to drink a plate of hot turtle soup. This was followed by a hearty open-air lunch. Coming back for tea he helped himself to poached eggs and preserved ginger, together with scones, hot cakes, cold cakes and Scottish shortcake. This was his preparation for dinner at 8:30 p.m. consisting of twelve courses, with champagne and cognac."

Whether the meal is a large or a small one, the stomach dutifully proceeds to churn it, homogenizing it and forcing out the excess air with an occasional belch. More important, the stomach now releases the gastric juices that liquefy the mixture and markedly change its chemical make-up. The juices consist mainly of an enzyme, pepsin, which starts the digestion of meats, and the very corrosive chemical hydrochloric acid, a compound so powerful, even in dilute solution, that it dissolves mortar off bricks. "We would have to boil our food in strong acids at 212° Fahrenheit," commented Dr. Benjamin Miller, "to do with cookery what the stomach and intestines do at the body's normal temperature of

Which way to save a choking victim?

More than 3,000 Americans choke to death each year. How to prevent such deaths has been a medical dispute since 1974, when Dr. Henry J. Heimlich *(below)* introduced the so-called Heimlich maneuver. It calls for powerful thrusts of a fist to the abdomen, forcing air into the windpipe to expel any foreign object. The Heimlich maneuver violates the slap-on-the-back technique favored for more than 30 years by the American Red Cross. In 1976, the Red Cross accepted the Heimlich maneuver but still advocated the back slap as a first resort—a step, contends Dr. Heimlich, that wastes precious time and that might lodge the object deeper in the throat.

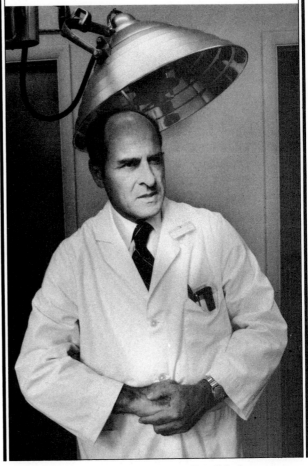

Cincinnati chest specialist Dr. Henry J. Heimlich demonstrates his lifesaving maneuver on himself. Placing his right fist above the navel but below the rib cage, he presses upward against his abdomen with his free hand and fist. A choking victim can use the technique alone, but it is more likely to succeed if performed by a trained rescuer hugging the victim from behind.

98.6°.'' The existence of such an extremely corrosive substance in the stomach and its role in digestion were confirmed more than a century and a half ago in a lucky experiment that has become a classic in the annals of medicine.

The man with a lid on his stomach

One morning in June of 1822, a U.S. Army surgeon, Dr. William Beaumont, was called to treat a young French-Canadian, Alexis St. Martin, whose stomach had been ripped open by an accidental blast from a shotgun. Dr. Beaumont did his best to pick out the shot and close the wound, but it was impossible; the edges, seared by powder burns, failed to seal shut. Miraculously, St. Martin survived, but with a permanent hole in his gut. Scar tissue formed around the opening so that it could be kept closed, enabling St. Martin to eat and digest normally. For some time he continued to require regular medical care, and Dr. Beaumont took him into his home (and for years afterward continued to provide financial support).

Although the hole ordinarily was shut, a hole it remained; with St. Martin's permission, Dr. Beaumont seized this unparalleled opportunity for study. Dr. Beaumont could force a tube through the hole to look inside St. Martin's stomach, extract secretions and even inject substances for testing. He once lowered in an oyster on a string, pulling it out an hour later to find the oyster partially frayed. He reinserted it for

another half hour, then drew the string out and saw that the oyster had dissolved completely. Dr. Beaumont also found that St. Martin's stomach lining changed color after meals; it also turned color when he was angry—the first direct evidence that the digestive system reacts to emotional stress.

The experiments continued off and on for years, and though St. Martin was ambivalent about his role as a guinea pig—he came to blows with companions who referred to him as the man with ''a lid on his stomach''—the repeated testing seemed to do him no harm. He went on to sire 17 children, and finally died at 86, more than half a century after the shotgun blast that made him famous in medical history. Meanwhile, Dr. Beaumont's work set an example for later direct observations of the stomach's interior, and confirmed once and for all that powerful acid secretions are an important part of digestion.

Among hydrochloric acid's critical functions is the destruction of harmful bacteria in the food. But it makes its most valuable contribution to digestion by adjusting the acidity of stomach liquids to the level required by the enzyme pepsin. Once this level has been attained, pepsin can begin to disassemble large molecules of protein, which are the main components of meat, fish and eggs, and are also important parts of beans and grains. Proteins, which provide the basic structural materials of all bodily tissues and all hormones and enzymes, are made up of chemical building blocks called amino acids. Protein molecules are too big to be absorbed into the bloodstream for immediate use in the body; they must first be broken down into their component amino acids. That is what the pepsin starts to do.

Babies have another enzyme in their gastric mix—rennin—which acts on milk, clotting it so that it will move more slowly through the system and allow the other digestive juices to extract its nutrients. The same coagulated solids form in sour milk, making Little Miss Muffet's ''curds''; the remaining clear liquid is the ''whey.''

That the high-powered, acidic mix in the stomach does its digestive duty so briskly without eating through the stomach lining is one of the system's greatest wonders, and one of its mysteries as well. Protection is provided by the lining of

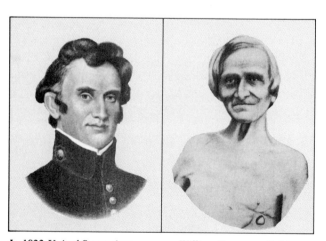

In 1822 United States Army surgeon William Beaumont (left) met his famous patient, Alexis St. Martin, shown at right in a picture taken more than half a century later, when St. Martin was 81. A shotgun wound had left a permanent peephole in St. Martin's stomach, allowing Dr. Beaumont to make the first direct observations of the digestive system at work.

the stomach wall, the mucosa, which constantly renews the mucous covering—replacing some types of cells every three days. Under normal circumstances, the mucosa bars the powerful juices successfully. But sometimes it weakens, permitting the acid to attack the layers beneath. The result is pain and, if the situation is not corrected, an eaten-away spot called an ulcer. The mystery is that ulcers occur in some people but not in others who have seemingly identical characteristics.

While the stomach lining repels hydrochloric acid, it absorbs a few other substances, most notably alcohol. That is why a martini on an empty stomach hits so hard—not enough real food has been absorbed into the bloodstream to dilute the alcohol that the stomach lining picks up. Carbohydrates, because they arrive in the stomach with their own enzyme, amylase, furnished by the saliva in the mouth, tend to be broken down faster in the stomach than protein. But hydrochloric acid quickly stops the amylase action, and the carbohydrates are sent along to the intestines. That is why a meal composed mainly of carbohydrates—such as Chinese food—soon leaves you hungry again.

Throughout these mixing and mashing actions, another muscular ring similar to the esophageal sphincter stands guard at the far end of the stomach. Called the pyloric sphincter—*pyloros* is the Greek word for "gatekeeper"—it permits the passage of food that has attained the right consistency and chemical composition. Over and over, the stomach's contractions fling partly digested food against the pyloric sphincter. But only when the food has been liquefied, made into a soupy, homogeneous mix called chyme, does the portal respond. Then the sphincter opens and allows small squirts of chyme to enter the intestines at a steady, continuous rate, some three squirts a minute.

The stomach's activity usually lasts about an hour; in another hour or two the stomach will have become quite empty, its chyme satisfactorily accepted into the intestine. A fatty meal leaves the stomach more slowly because the intestines are slow to handle fats and they signal the stomach to delay release; this is why fatty meals, unlike carbohydrate meals, leave a feeling of fullness that lasts for hours. As the stomach empties, it shrinks in size and waits for the next meal, little more than air remaining in it. If the next meal is long in coming, the stomach muscles will contract expectantly—hunger pangs. During its several hours of activity, the stomach has delivered only a very small part of the meal's nutrients into the bloodstream. In Dr. Miller's figure of speech, the food is still on the table. Full absorption—the life-sustaining purpose of digestion—takes place only after food leaves the stomach and enters the intestines.

Why you need a liver

The intestines depend on the stomach to provide them with food suitably mashed up and acidified. They then add some digestive chemicals of their own manufacture. But they also make use of chemicals and processing provided by three independent organs: the liver, the pancreas and the kidneys. These subsidiary units fill many roles in the body. For digestion they manufacture enzymes and other substances to be funneled into the intestines, or they dispose of wastes from the processing reactions, or they regulate the supply of essential materials—or they do all these things at once.

The liver helps to digest only fats, but the pancreas—about six inches long and just behind the stomach—is crucial to the digestion of all kinds of food. Perhaps its best-known job is to make insulin, which the body uses in converting the products of digestion into the energy of life; when the pancreas fails in this duty, the result is diabetes mellitus.

Its role in digestion is equally vital, for it secretes an alkaline juice, rich in enzymes, that flows into the intestines just below the pyloric sphincter. One of the enzymes provided by the pancreas is amylase (also found in saliva), which continues the process of splitting starch molecules into absorbable sugars. Another enzyme, pancreatic lipase, furthers digestion of fats, changing them to fatty acids and a clear, syrupy liquid called glycerol (chemically the same as the glycerin sold in drugstores). Two more enzymes, trypsin and chymotrypsin, join pepsin in attacking the proteins. Pancreatic juice is called into action by hormones that are released from the intestinal wall just as the chyme begins coming out of the stomach through the pyloric sphincter. In one day the pancreas produces more than a quart of its enzyme-filled juice.

Cooperating with the pancreas to break down fats and neutralize acids is the outflow from its neighboring organ, the liver. Not only is the liver the largest internal organ—weighing three and a half pounds and almost as big as a football—but it is perhaps the most sophisticated. Ancient scholars, unwittingly honoring its critical role in maintaining the body's chemistry, called it the seat of the soul and the very center of life; in Babylonian sacrificial rites, kings scrutinized animal livers for possible messages from their gods.

The liver, Dr. Miller noted, is the body's "fuel storage and supply office, housekeeper, and poison control center." Among its functions are prevention of disease (it filters out and destroys bacteria); production of proteins needed for blood clotting; neutralization of drugs (it combines them with other chemicals so that they may be safely eliminated); storage and release, as needed, of the energy-supplying sugar glucose, received from digestion; and conversion of proteins and carbohydrates into fat for storage in the tissues. Not the least of its duties is the making of bile, a liquid that serves mainly as a detergent. Fats, for complete digestion, must be emulsified—broken up into water-soluble globules that will mix readily with the pancreatic enzyme lipase. Chemical salts in the bile emulsify fats and also ferry the digested nutrients over to the intestinal wall for absorption.

Every day the liver produces about a quart of bile, which is yellow-green or golden brown when first secreted—depending on what waste products the liver is handling—but may change color slightly as it combines with other substances in the digestive tract. Another word for bile is gall, an Old English term for yellow; the Greeks called it *chole* and said that someone who generated too much yellow bile was "choleric." (The Greeks also believed, erroneously, that some bile was black, and that an excess of it caused an individual to be "melancholic.") To be "bilious" is to be ill-tempered; Hamlet bemoaned his fate, "It cannot be but I am pigeon-livered, and lack gall to make oppression bitter."

As the bile flows down the so-called common duct leading from the liver to the small intestine, it is shunted aside at a junction to a special storage bag called the gall bladder, just beneath the liver. There it is retained in concentrated form, to be squeezed out after mealtime, as soon as fats leave the stomach and enter the intestines. When the bile is released, it returns to the common duct. Just above where the passageway connects to the small intestine, the duct also receives the alkaline juices from the pancreas, enabling the pancreatic enzymes trypsin and chymotrypsin to go to work on proteins immediately. The fat-digesting lipase, however, can complete its job only after bile emulsifies the fats.

Some of the bile ingredients may solidify into small lumps to become what are known as gallstones. One or more stones may exist harmlessly in the gall bladder for years, but if one exits and begins blocking the passages to the small intestine, there ensues the yellow skin of jaundice as bile pigments enter the blood; severe pain and digestive upset may also occur because the supply of bile is cut off. The conventional remedy is surgical removal of the gall bladder, a simple operation that generally has no ill effect—storage and concentration of bile in the gall bladder does not seem to be necessary for digestion. New medicines, however, can dissolve some gallstones, eliminating the need for surgery in certain cases. Now and then the liver may succumb to other, more serious disorders, and a frequent telltale sign is jaundice, indicating the inflammation of hepatitis or even the permanent liver damage called cirrhosis.

As part of its housekeeping duties, the liver rids the body of certain harmful by-products of protein processing. The chemical actions that break down the amino acids of protein and rearrange them into other materials generate as a by-product poisonous ammonia. The liver is largely responsible for changing ammonia into urea, which can be excreted. The urea filters into the bloodstream for delivery to the body's two most essential organs of elimination—the kidneys.

Perhaps because kidneys perform such a multitude of life-sustaining functions, every person has two of them although one can handle the job alone—insurance in case one fails through accident or disease. The kidneys are lodged on either side of the backbone at the lower edge of the rib cage. Like the liver, they remove toxins from the blood; they also manufacture an enzyme regulating blood pressure, and monitor

The pipeline that supplies the body

The body gets everything it needs except air from a tube that can be likened to the hole in a doughnut; it is not so much a part of the body as a passage through it. This inner space, the digestive tract, is nevertheless a complete biochemical world, serving partly as a pipeline and partly as a vat in which bubbling, churning reactions convert food into compounds that the body can use.

The tract comes with its own skin, a mucous membrane that is an inner counterpart to the body's outer covering of epidermis—the cells of the two skins are remarkably similar. Each of the tract's five major compartments—mouth, esophagus, stomach, small intestine and large intestine (diagramed at right)—has its own microclimate, subject to shifts in pressure, variations in humidity, and cónstant changes in acidity.

These compartments are surrounded by components that help the tract accomplish its work with 95 per cent efficiency—only 5 per cent of the food eaten is eliminated as waste. Organs such as the liver supply fluids to the tract through ducts, supplementing the juices that are secreted by the membrane inside the tract itself. Nerves bring operating instructions from the brain.

The body gets the supplies it needs after the process of digesting, or breaking down, the food is completed; only then are the nutrients absorbed through the permeable membrane into the blood and lymph systems. Even the water left in the intestines' end products is not wasted; most of it is absorbed and eventually recycled by the kidneys.

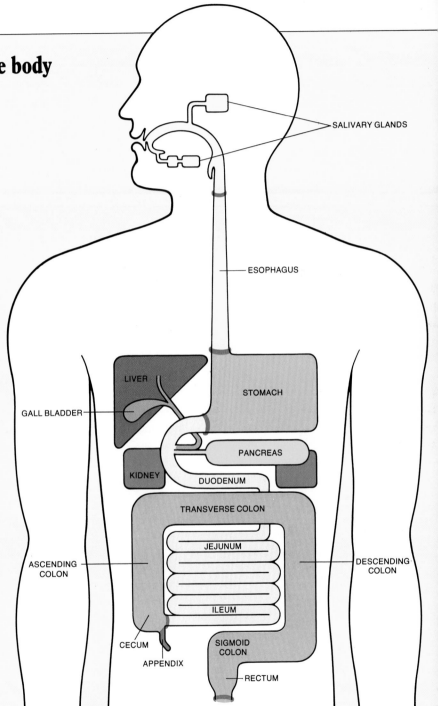

A MAP OF THE TRACT
Sphincters (red) control the passage of food broken down by fluids from the salivary glands, stomach, pancreas, liver and gall bladder, and the duodenum and jejunum of the small intestine (light pink). The ileum and large intestine (dark pink) extract water —processed in the kidneys. The appendix serves no purpose.

Digestion—partly chemical, partly mechanical

Before nutrients can be absorbed *(overleaf)*, food particles must be digested—converted into usable compounds by acids and enzymes that are released as the food moves from point to point along the tract. Different kinds of food are treated in different parts of the system *(opposite)*—some sugars are partially digested in the mouth, while proteins begin to be broken down in the stomach; most of the digestion and absorption, however, takes place in the small intestine. Passage from one section to another is con-

trolled by the one-way valves known as sphincters *(below, right)*.

To move the food along this processing line, the esophagus *(below, left)*, stomach and intestines employ a special type of muscle action, peristalsis. Their muscles, responding to nerves that sense the presence of food, contract to squeeze the tube shut as rapidly as 12 times a minute. These contractions move in waves down the tube, squeezing the food ahead in much the same way fingers squeeze toothpaste out of a tube.

A squeeze play to move food through

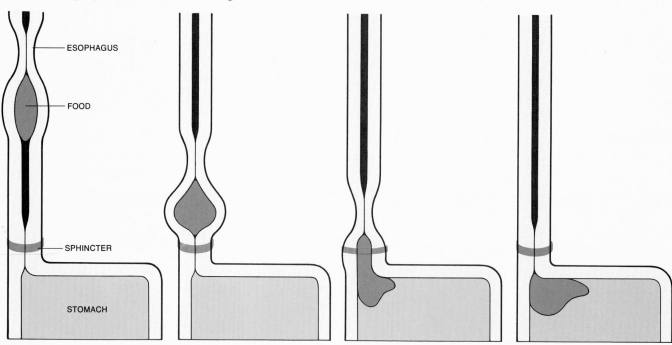

ESOPHAGUS

FOOD

SPHINCTER

STOMACH

PERISTALSIS: A WAVE OF CONTRACTION AND EXPANSION
When food, chewed and swallowed, enters the esophagus, it is squeezed down by peristalsis. A section of muscle contracts (above, left) as a section ahead relaxes to make room for the food being pushed down. As successive sections expand and then contract, the food is pushed until it reaches the esophageal sphincter (above, right).

A VALVE TO KEEP FOOD WHERE IT BELONGS
Triggered by the wave of relaxation preceding the food, the circular muscle of the esophageal sphincter, at the bottom of the esophagus, also relaxes (above, left), permitting food to enter the stomach. The sphincter then immediately constricts to its normally closed position (above, right), preventing acidic stomach juices from backing up.

Where food becomes nutrients

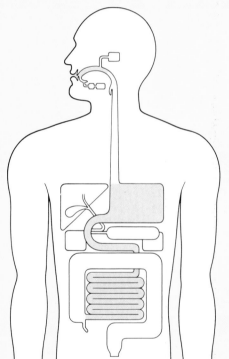

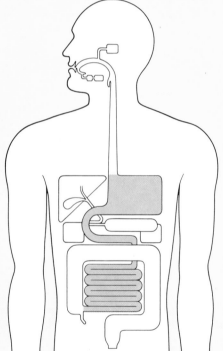

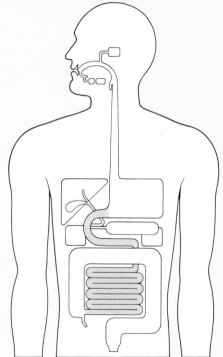

CARBOHYDRATES: A BIG JOB
Carbohydrate digestion (light purple) begins in the mouth, where the enzyme amylase in saliva breaks down a few starches. The process continues in the stomach and is completed in the small intestine, yielding absorbable sugars—mostly glucose.

SPLITTING PROTEINS IN THE STOMACH
Protein digestion (green) is partly accomplished in the stomach, where the enzyme pepsin, activated by hydrochloric acid, breaks up these large molecules. In the small intestine other enzymes finish splitting protein into nutrients.

THE SMALL INTESTINE'S SPECIALTY: FATS
Fat digestion (blue) takes place entirely in the small intestine, where bile salts make fats soluble in water; then the enzyme lipase splits fat droplets into absorbable glycerol and fatty acids. Here, too, almost all other nutrients are absorbed.

Sieves to screen the nutrients

Digestion converts food molecules, many of them very large and insoluble in water, into small, soluble molecules that can be picked up by body fluids and transported where they are needed. The membrane lining the small intestine and colon, in addition to secreting mucus and digestive juices, also serves as a selective filter; it is covered with special absorptive cells that sample the various nutrients and fluids and admit them at appropriate points.

Most of this absorption takes place in the small intestine, where the fiber-like villi *(below, left)* are coated by absorptive cells. The bulk of the fats are transferred by the villi into the lymph system, the other nutrients into the bloodstream, for delivery to the muscles, glands and brain. Then the kidneys *(below, right)* take over the filtering function, monitoring and regulating the chemical balance of fluids in the body. Each day as much as two quarts of surplus water and toxic wastes is flushed from the blood by these vital sieves.

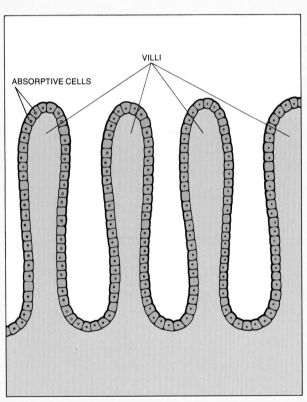

VILLI TO TAKE IN NUTRIENTS
Absorptive cells coat the intestinal projections called villi. These cells are formed in the depressions between villi, then migrate to the tips of the villi during their five-day life spans, after which they are shed and replaced. Through a fuzzy border of even tinier microvilli (not shown) on the surfaces of the cells, nutrients are admitted to the blood and lymph vessels.

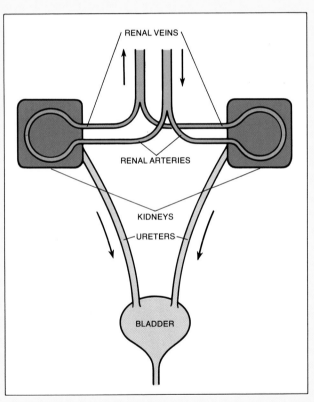

KIDNEYS TO RECYCLE FLUIDS
In the kidneys, blood fluid from the renal arteries passes through 2.4 million tiny filters called nephrons; then, cleansed, it is returned to circulation through the renal veins. Nephrons remove excess sodium and potassium, as well as urea, a by-product of the liver's food processing. The wastes, dissolved in water to form urine, are channeled through ureters to the bladder.

and maintain the body's chemical balance. And while not actively engaged on digestion's front lines, as processors of its end products they are linked closely to it.

Each kidney, about the size and shape of a fist, is a sophisticated filter. Its 1.2 million tiny operating units, called nephrons, continuously extract fluid from the blood, remove offending substances, and return the purified liquid to circulation. Some 200 quarts of water and dissolved chemicals pass through the nephrons daily, more than twice the body weight of an average adult. Over 99 per cent of this fluid is sent back into the blood, but one or two quarts, containing urea and other chemicals, are removed as urine. As it accumulates, the urine is forced by peristalsis down through a pair of tubes to the bladder for disposal.

In handling this flow of liquid, the kidneys perform one of their most essential tasks—regulating the body's water content: Almost 60 per cent of body tissue is nothing else. The blood is nearly all water—80 per cent. Water seeps from the blood into individual body cells and back again, carrying nutrients, hormones and waste products. One factor controlling this two-way passage is the balance in the body of the element sodium, a component of common table salt. The kidneys filter and extract sodium, and thus regulate the amount of fluid in the tissues; too much sodium, for example, causes the body to retain more water than it should.

If the kidneys stop working, the consequences are grim. Toxins accumulate in the blood; tissues bloat with water. Unless kidney action can be restored, death follows within a few days. Fortunately, the kidneys are hardy—and duplicate—and such crises seldom occur. A far more common kidney disorder is much like gall bladder trouble: kidney stones. Treatment varies with the kind, size and location of the stones: drugs to dissolve them, surgery to remove them or, if they are still small, the imbibing of great quantities of water to flush them out. There is seldom lasting damage.

Where the fuel gets into the body

The kidneys, liver and pancreas are essential to the smooth functioning of the digestive system. But they only help to make the main part of the system, the intestines, perform the vital job of disassembling food molecules and converting them into compounds that the rest of the body can use for repair, for fuel and for defense against disease.

Most of this chemical processing goes on in the small intestine, so called not because of its length, for it is prodigiously long—as much as 22 feet in adults, coiling round and round in the abdomen—but because of its narrowness. Most of it is an inch and a half in diameter. That distinguishes it from the tract's last major segment, the large intestine, which is far shorter but up to two and a half inches wide; together the small and large intestines constitute the bowel.

Small though the upper bowel section may be, its role is crucial. An individual who for some reason lost the use of mouth, esophagus and stomach could still acquire nourishment. Furthermore, the large intestine could be dispensed with if necessary: The small intestine alone would sustain life. But anyone who lost most of the small intestine probably would have to be fed intravenously *(pages 122-127)*.

The small intestine is a kind of pipeline inside which digestive processes take place, using materials supplied mainly by other organs but also partly by cells of the intestinal lining. It also is the main absorber of nutrients into the body. Its huge length extends its inner surface for a maximum intake. The surface is increased by the complex interior lining of the small intestine. Looking something like a crumpled Turkish towel, its folded lining contains a mossy nap of as many as 60 million tiny projections, called villi. And the villi are covered with tinier projections of their own, thousands of microvilli, each about .00004 inch long. Together the folds, villi and microvilli multiply the absorptive area 600-fold—to a total of about 300 square yards, the size of a tennis court.

The velvety lining of the small intestine does far more than absorb like a towel. It plays an active role in digesting food. Cells covered with microvilli travel out along the shafts of the villi, until they reach the tips and are sloughed off, carrying digestive enzymes with them to add to those already in the chyme. Some of these enzymes continue the transformation of fats into glycerol and fatty acids, while others complete the breakdown of protein into amino acids. Still others go to work on the remaining carbohydrates to produce glucose, galactose and fructose—the so-called simple sugars. In most people an enzyme called lactase processes lactose, the sugar in milk products; but many other people—including an estimated 70 per cent of black Americans—do not have this enzyme, and they may get cramps and diarrhea if they consume large amounts of milk or cheese.

To ensure that the villi extract and pick up every nutrient possible, muscles in the intestine rock the chyme mixture back and forth, and the villi themselves are similarly in constant movement, swaying from side to side and up and down.

The villi begin to work their magic in the first section of the small intestine—the duodenum—a C-shaped tube some 10 inches long that leads away from the pyloric sphincter. That it should bear a Latin name meaning "twelve" when it is neither 12 inches nor 12 centimeters in length is due to its having been measured by early anatomists in finger widths. It is here that the chyme, which is very acidic upon leaving the stomach, must be neutralized so that the enzymes can work. And acid would eat away the lining of the small intestine, which is less resistant than that of the stomach. In most instances the neutralization occurs satisfactorily, but now and then—again for reasons not clear—the acid wins out and the lining of the duodenum acquires an ulcer. Duodenal ulcers, in fact, are the commonest form of the malady.

From the duodenum, the soupy mix of chyme and enzymes moves by peristalsis into the small intestine's two remaining sections, the jejunum and the ileum. Here, as more nutrients are absorbed, others receive further treatment. By the time the chyme mixture leaves the ileum, most digestion is completed. Nearly all foods have been converted into compounds the body can use, and the compounds have been absorbed by the microvilli—most of them go directly into the bloodstream, but the bulk of the fats are put into a complementary distribution network, the lymph system.

The small intestine accomplishes its life-supplying work in three to ten hours; at the end of that time, the remains move on to the last major part of the alimentary canal, the large intestine. Its function is quite different from its predecessor's—a fact doctors recognize in their terminology. The entire digestive system up to this point—including esopha-

gus, stomach and small intestine—they call the upper gastrointestinal (GI) tract. The large intestine, all by itself, is called the lower GI tract.

Capturing the most important ingredient: water

The mission of the large intestine is essentially to take the remnants of the digestive process, drain them of water, move them toward the exit portal and get them ready for ejection. It does so via a five-foot-long tube that—unlike the convoluted small bowel—follows a simple course. Starting in the lower right-hand corner of the abdomen, it travels up the right side, across the body and then down.

The entrance to this upside-down U-shaped vestibule is guarded by another of those internal gates that keep materials from going astray. In this case the ileocecal valve—the division between the ileum and the large intestine—keeps the large intestine's contents, which become laden with bacteria, from backing into the relatively sterile confines of the small intestine. Waste goes through this valve and into the cecum, the first section of the large intestine, a wide area that extends several inches downward from the entrance valve. The cecum thus forms a pouch where processed chyme from the small intestine collects before moving up one arm of the second section of the large intestine, the colon.

At the bottom of the cecum there appears a mystery, in the form of a small blind alley four inches long—the appendix. It has no apparent function; most likely it is simply a relic of human evolution. Occasionally the appendix will trap waste materials, becoming irritated and then inflamed. This is appendicitis, signaled by a sharp pain felt first in the pit of the stomach; doctors then almost always cut out the relic.

The chyme collecting in the cecum moves gradually up the ascending arm of the colon, works its way across the transverse section, and then travels down the descending arm. It takes a final S-curve through the so-called sigmoid colon and arrives at the rectum, a chamber five inches long that is the last stop on the digestive journey. The rectum is held closed against the outer world only by the final muscular ring controlling the anus.

During this transit the chyme undergoes major physical

modification. By now it consists partly of cellulose, the indigestible roughage left over from vegetable matter; it also contains a number of other substances that the body does not need, plus remnants of the digestive-tract lining that have been sloughed off in the unending renewal process, and lastly the bile pigment. And there is plenty of water, which is more essential to life than any of the other food components ordinarily considered nutrients. Most of the bodily supply of this vital substance comes from drinking liquids, but about a third comes from eating solid food—a cooked hamburger is 55 per cent water, a boiled potato 80 per cent. The water is removed from the residue in the colon, whose walls soak up more than 90 per cent of the liquid—more than a quart a day—drying out the chyme.

An important change that happens to chyme in the colon is the proliferation of bacteria, which were largely killed by acid in the stomach. As the mix steadily grows more alkaline in the small intestine, they begin to multiply, particularly in the ileum. In the colon these surviving bacteria multiply furiously. Far from being universally harmful, the colon's bacteria are mainly benign and perform a valuable service in breaking down remaining proteins and fats. They also make vitamins needed by the body, chiefly vitamin K, which is absorbed by the colon. But some of the bacteria can be dangerous, and home plumbing and sewage systems must be carefully built to prevent the spread of germs.

A colony of busy, benign bacteria

The bacterial action, basically a fermenting process, yields a combination of hydrogen, carbon dioxide and methane that makes the colon's contents gaseous to varying degrees, and this can bring discomfort. Bacteria also cause the odor of the material that has now resulted from these transformations— the so-called feces (meaning dregs), or stool.

Passage through the large intestine is much slower than in the previous sections of the gastrointestinal tract. Material can remain there about 24 hours and often much longer, with no harm done. Peristalsis here is of a different nature. Instead of being continuous, it comes in large waves three or four times a day that move the contents along swiftly to the next

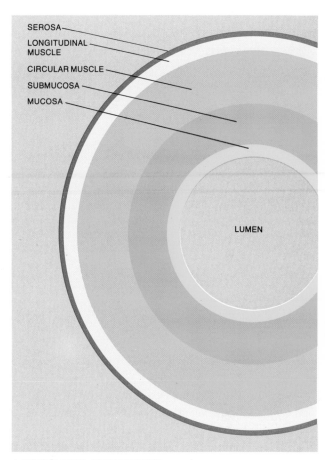

A MULTILAYER, FLEXIBLE PIPE

The layers of tissue making up the wall of the 30-foot-long digestive canal are represented in this cross-sectional diagram. Proceeding outward from the lumen, or central cavity, is first the mucosa, which in the small intestine is coated with finger-like villi to absorb digested food; next is the submucosa, which connects the mucosal tissue to two layers of muscle— one circular and the other running lengthwise—whose contractions move food through the canal. Surrounding the whole is the serosa, a gristly protective sheath.

This marble relief from the tomb of Jason, an Athenian physician of the Second Century A.D., depicts the simplest and probably the oldest form of examination for digestive ills—gentle probing of the abdomen. The domed object beside the patient is a giant cupping glass, a suction device that was placed on the skin to draw blood to the surface and away from an inner source of pain.

section. One stimulus producing such activity is the entry of food into the stomach, which causes signals to be sent ahead to make way in the intestines for the new arrivals. Whatever material has previously lodged in the sigmoid colon may thereupon be pushed into the otherwise empty rectum, which in turn sends a signal to the brain announcing that it is full and ready to be cleared out.

Sometimes, as it happens, the rectum fills too often—or seemingly not often enough. It is such variations in accustomed bowel rhythm that prompt much of the concern people have about their digestive systems.

What may appear to be abnormalities in the type and frequency of bowel movement are usually brought about by the extent to which the feces have been relieved of water in the large intestine. Material that passes through quickly, perhaps because it contains toxins that the body hurries to expel, spends too little time in the drying process and emerges in a watery state, as diarrhea. Material that passes through too slowly for one reason or another may become unduly parched, so that it then moves along with even greater difficulty, sometimes resulting in outright constipation. But slow-moving feces can be perfectly natural and may not signify anything at all wrong.

Individuals differ in what can reasonably be considered a normal frequency; for some, it is several times a day, while for others it may be only once every few days—and either pattern may be healthy. As a consequence, much of the nonprescription medication sold to promote ''regularity'' is not necessary. Used to excess, it may even do harm.

Worry over the rhythm of operation of the hard-working large intestine may be largely unjustified. But it is subject to other ills that can be serious. The large intestine is particularly sensitive to emotional stress. Its commonest reaction is a malady known as irritable bowel syndrome or spastic colon, which causes abdominal pain and either constipation or, less often, diarrhea. Untold numbers of people suffer from it off and on for years—and its permanently successful treatment is often difficult to effect. It causes as much time lost from work as the common cold.

More serious are diverticulitis, in which pouchlike extrusions that have developed in the intestinal wall become infected; and a group of baffling inflammatory disorders called ulcerative colitis and Crohn's disease.

To the rescue: new drugs, new instruments

Exact causes are known for only a few of the ailments of the digestive tract. Yet more and more of them are now successfully treated or even prevented. Broadened knowledge about digestion has helped. So have new drugs, such as cimetidine and sucralfate for gastric ulcers, and chenic acid for gallstones. But much of the credit must go to improved diagnosis, made possible by the invention of several remarkable devices for examining the tract *(pages 92-105).*

One new instrument that greatly aids diagnosis, and can sometimes be used for treatment, is the fiber-optic endoscope, invented at the University of Michigan. A flexible viewing tube about half an inch in diameter and up to six feet long, it contains thousands of tiny, precisely aligned glass strands that carry clear images from the probing end to an eyepiece or camera at the other end. Inserted through the mouth, it can be maneuvered down the esophagus, through the stomach and even into the duodenum. Another type, the colonoscope, can be slid up through the rectum to probe the length of the large intestine, and even the ileum.

With an endoscope, the doctor can inspect almost every square inch of interior surface—examining inflammation, noting the exact location of ulcers and even identifying the cause of an obstruction. He can take photographs or transmit television images to a screen, inject opaque fluids to make clear X-rays or, with the use of tiny forceps or snares attached to this instrument, remove samples of possibly harmful growth. The scope can even be used to extract swallowed objects like beads and safety pins.

The value of such clear views of the digestive organs is indicated by the case of a woman who had been hurt in an automobile accident. Her severe pain was diagnosed as pancreatitis—inflammation of the pancreas. Exploratory surgery revealed no cause for her disorder, but the pain was so great she had to be hospitalized five times in the following six years. Finally doctors used an endoscope to inject a dye

directly into the duct leading from the pancreas. This allowed them to take an especially clear X-ray, which showed for the first time a slight injury at the tip of the woman's pancreas, presumably from the accident. Surgeons excised the offending part and she was cured.

Two other new diagnostic tools have been of enormous benefit. One involves ultrasonics—using very high frequency sound waves to obtain images of body structures—and it is particularly useful in finding gallstones, which often fail to show up in an X-ray. The other is the CAT scan—computerized axial tomography—in which a computerized revolving X-ray camera gives cross-section views of a patient's insides that can be far more informative than conventional X-rays.

Even more welcome than improvements in professional diagnosis and treatment of digestive disorders is new understanding of ways to prevent these ills on your own. A few ills have finally been traced to specific compounds in foods—the lactose in milk is one—and such disorders can be eliminated simply by not eating the offending substance. Many other ideas about diet have been modified. Ulcers were long blamed on spicy and acidic foods. Doctors now believe that diet is neither a cause nor a cure—the benefits of the rich, milky meals once prescribed are today questioned, and some experts think milk may be harmful.

An even more striking reversal of expert opinion has altered advice on fibrous foods—vegetables and grains that contain indigestible roughage. During the 19th Century, fiber was hailed as a cure-all. Then, early in the 20th Century, it came to be considered a cause of gastrointestinal disorders. Now most doctors have changed their minds once again and believe—mainly on the basis of circumstantial evidence—that eating lots of fiber is necessary to a healthy digestion.

Most of the hard-won understanding of digestion confirms the value of care and moderation: limiting the use of alcohol and other drugs, watching out for contaminated food, controlling stress through exercise and rest, eating what agrees with you and avoiding what does not. Such commonsense measures will make Mr. Polly's upheavals, rumblings and thunderings just amusing reading—and keep them from becoming personal experience.

The system's hidden wonders

Over a normal life span, some 90,000 pounds of solid food and 55,000 quarts of fluid course through the human digestive tract, a fragile, membrane-lined passage that meanders some 30 feet through the torso. In the tract, nutrients are extracted, processed and absorbed in a series of complex mechanical and chemical actions. The intricacy and eerie beauty of the system have been captured by Swedish photographer Lennart Nilsson, who used specially built optical and electronic devices to take the photographs on these and following pages.

Nilsson's pictures, taken of cadaver tissues and organs, provide rare interior views of the funnels, valves, tubing, reservoirs and mixing tanks forming the mouth (below and right), esophagus, stomach and intestines (following pages). In a normal 48-hour passage through this refinery, foods and fluids move at varying speeds—an inch per second through the esophagus, an inch in 24 minutes in the large intestine. Food particles are subjected to some 20 chemical agents, many of them extremely potent.

The digestive system is self-regulating and very efficient. Automatic contractions of intestinal muscles provide a conveyer-belt action that shoves food from one processing station to the next. The stomach's 35 million gastric glands, aided by the nearby liver and pancreas, produce a gallon of digestive fluids a day. The small intestine sends most of the nutrients into the blood and lymph systems, the body's delivery networks; the large intestine extracts the last bit of food value and wrings water from the residue for recycling. The system even repairs itself. Every three to five days, worn-out surface cells lining the stomach and intestines are sloughed off and replaced in an unending renewal process.

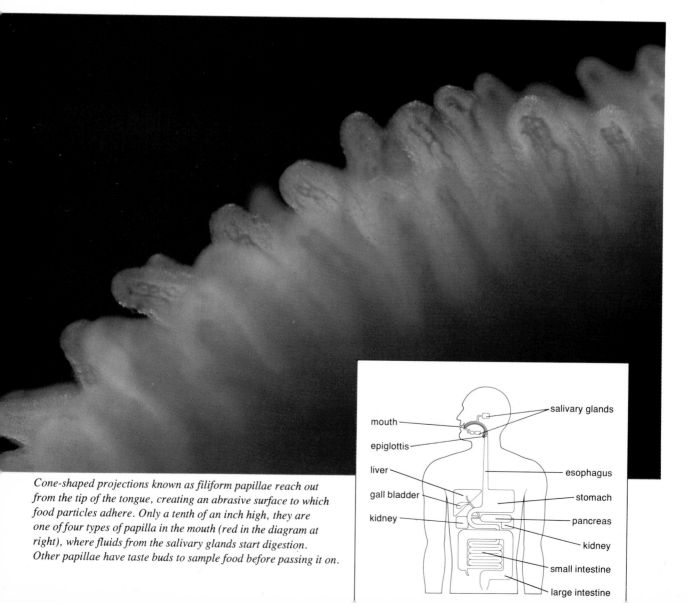

Cone-shaped projections known as filiform papillae reach out from the tip of the tongue, creating an abrasive surface to which food particles adhere. Only a tenth of an inch high, they are one of four types of papilla in the mouth (red in the diagram at right), where fluids from the salivary glands start digestion. Other papillae have taste buds to sample food before passing it on.

salivary glands

mouth

epiglottis

liver

gall bladder

kidney

esophagus

stomach

pancreas

kidney

small intestine

large intestine

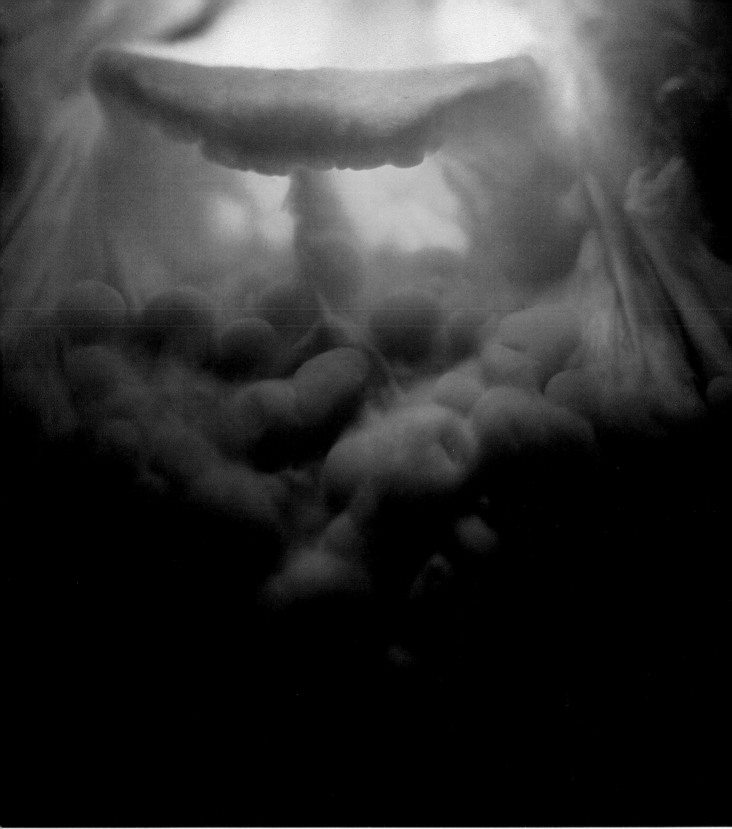

The epiglottis, a leaf-shaped flap of cartilage, stands guard over the entrance to the windpipe, permitting air to flow in during breathing, as above, and snapping shut to keep food from the windpipe and lungs during swallowing. Food as it is swallowed slides over the closed flap and into the esophagus, a 10-inch channel to the stomach directly behind the windpipe.

PHOTOGRAPHS BY LENNART NILSSON

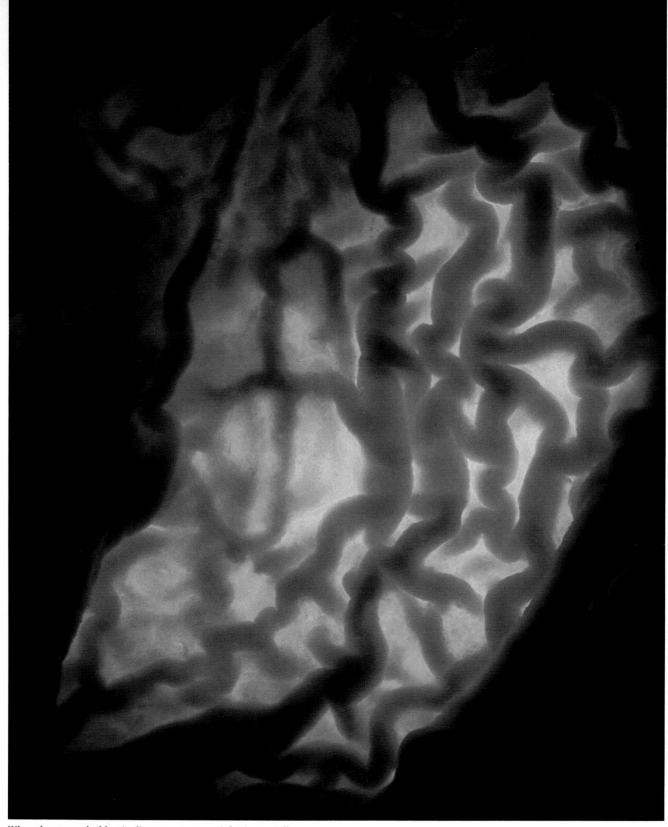

When the stomach (blue in diagram at upper right) is partially
empty, as in the picture above, the membrane lining its surface
falls into loose, elongated folds, like a deflating balloon;
during digestion, the folds stretch out and the stomach inflates into
a J-shaped bag a foot long and five inches in diameter, and
capable of accommodating two to three pints of food.

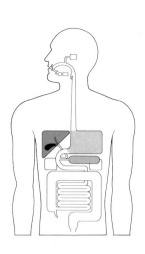

Viewed from underneath in the photograph at right, the liver (red in diagram) seems to envelop the gall bladder (black), which nestles between the left and right lobes of the liver. The gall bladder stores and concentrates bile, the digestive fluid manufactured by the liver.

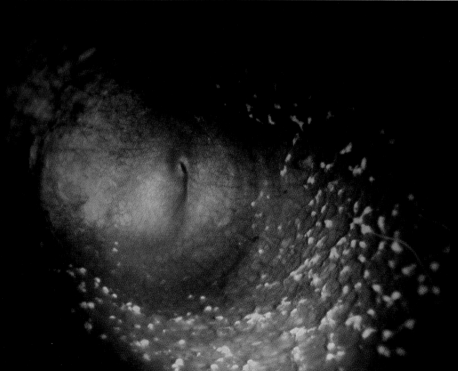

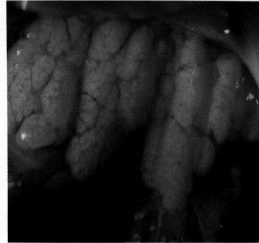

A close-up of the pancreas (green in diagram at top left) reveals small, blood-rich lobes composed of sacs called acini—the Greek word for ''grapes'' — whose cells produce fluids for the digestive tract. Other cells, which are located between the acini, secrete insulin, the vital hormone that regulates the level of sugar in the bloodstream.

Tinted green from pigments in the bile fluid it stores, the interior of this gall bladder is speckled with yellowish deposits of cholesterol, a component of bile. Although cholesterol plays no known role in digestion, it can cause excruciating pain if its deposits harden into a mass—a gallstone—that then blocks the flow of bile out through the bladder's duct (dark area, center).

The small intestine, shown in the photograph below and in red in the diagram at bottom, coils compactly in the abdomen. An inch and a half in diameter, it is 12 to 22 feet long and lined (right) with a thick carpet of fibers, providing an absorptive surface equal in area to the floor space of a large two-story house.

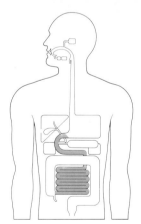

A view down the lumen, the interior of the intestinal pipe, reveals tiny absorptive projections called villi, silhouetted at the edges of the lumen's circular folds. The villi, .04 inch long, are in turn coated with smaller projections, the microvilli. Extending .00004 inch from the villi surfaces, these submicroscopic fingers do most of the processing and absorbing of nutrients.

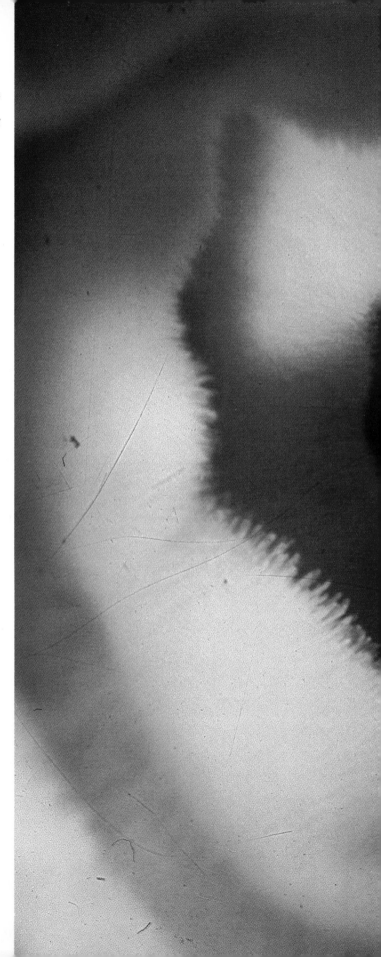

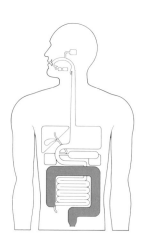

The smooth folds of the large intestine (red in diagram above) are speckled with special glands that in this photograph resemble skin pores. Each day cells in the glands extract from passing food residues more than a quart of water, which is recycled and eventually filtered and excreted by the kidneys. The large intestine also yields vitamin K, which is synthesized by huge colonies of benign bacteria as they break down indigestible roughage.

CHAPTER 2

When to go to the medicine cabinet

The best preventive—eat sensibly
Choosing the right antacid
Extra ingredients that may or may not help
How to guard against food poisoning
The traveler's bane
Motion sickness and morning sickness
Irregularity: an overrated problem

It happens to everyone from time to time. Suddenly, out of the blue, the stomach turns a somersault and the nauseating echo of last night's tuna fish starts rising, or perhaps a dull, insistent pain starts nagging at the abdomen, or inner rumbling and gurgling herald a bout of diarrhea. What has happened? Will it all go away in a few hours, or is this the start of something serious?

Such distress sends most people to the medicine cabinet to try some kind of stomach-soother; they pop, spoon, ladle, squeeze or pour unmeasurable quantities of nonprescription medicines into themselves in hope of relief from various ills—real or imagined—of the digestive system. Of the more than five billion dollars spent on nonprescription drugs each year in the United States, almost one fourth goes for substances intended to mollify one or another part of the gut.

In most cases the hope for relief is rewarded. But before you start dosing yourself you should try to figure out just what the trouble is. For distress in the digestive tract can arise from any number of underlying causes, some mild and transitory, others more serious and abiding. Nausea, for instance, can mean nothing at all or it can be a symptom of cancer, heart attack or acute depression. Furthermore, no drug is totally safe under every circumstance, and if the upset stomach signals the onset of major illness, the wrong medication may be downright harmful.

Perhaps 99 times out of 100, though, what feels like a digestive storm is a digestive storm and nothing worse. Most common is simple indigestion, brought on by some indiscre-

tion at the dinner table. Eating too much strains the stomach beyond its normal capacity and interferes with normal functioning. Or maybe the problem is what was eaten—fiery lamb curry, tasty but fierce, can play havoc with the digestion because highly spiced foods, while they may stimulate the gastric juices, can also irritate both stomach and intestines. Then too, either worry or excitement can bring on indigestion, since both often incite an outpouring of gastric acid that, in excess, will irritate the stomach lining. So will eating too fast. The rapid gobbling of a meal will catch the stomach by surprise, overloading it with food and swallowed air before it is ready to go to work.

Often stomach irritation is so pronounced that it goes beyond indigestion. The stomach lining becomes inflamed, a condition known by the umbrella term gastritis—or, if inflammation extends to the intestines, as gastroenteritis. Both are catch-all designations for a multitude of digestive ills; sometimes the trouble can be handled with home remedies, and sometimes not. One common source of inflammation is alcohol; in excess it attacks the stomach lining, bringing pain and nausea. Large quantities of aspirin have the same effect. Another cause of inflammation is so-called intestinal flu—various shadowy types of viral infection that assault the stomach and intestines. Food poisoning is yet another culprit, as bacteria and toxins from contaminated meat, fish, eggs or drinking water ravage the tract.

The symptoms of gastritis and gastroenteritis are generally similar to those of simple indigestion, but often painfully

Fizzing into bubbles in a glass of water, the effervescent tablet (right) is one of many antacids that Americans buy—at a cost of $500 million a year—to relieve the upsets that result from consuming too much—or the wrong kind of—food or drink.

intensified: nausea, gnawing stomach pains, perhaps vomiting as the system attempts to rid itself of the offending substance, and sometimes a bad case of diarrhea. Muscular spasms in the inflamed intestines may bring severe abdominal cramps. If the cause is bacteria or a virus, fever and chills may develop; extreme cases of either ailment require professional attention.

Less disturbing than any of these ills is a problem that nonetheless annoys millions of people: constipation. In a way it is a mirror image of stomach upset, for instead of everything happening, nothing does. Constipation is almost always treatable at home, but many people do so incorrectly, with medicine that actually hurts them and even perpetuates the condition.

The best preventive—eat sensibly

The best way to keep all these kinds of upset at bay is the simplest: Prevent them by eating sensibly. Follow a diet that includes plenty of liquids, some milk products, and a variety of fruit and vegetable carbohydrates to balance the proteins of meat and fish. Eat slowly. Be wary of dishes that are highly spiced. Garlic, horseradish, curry, chili peppers, pickles and vinegary salad dressings all can upset a sensitive mucosa. So can lobster, crab, various nuts and fried foods, all of which, because of their high fat content, are slow to digest. Strive for moderation in alcoholic beverages. Make sure kitchen utensils are kept clean, patronize markets where you know the food is fresh and, when you dine out, consider sanitary standards in choosing a restaurant.

When a digestive upset does strike, any one of four danger signs should send you quickly to the doctor. One is truly severe pain or discomfort. Another is successive episodes of vomiting, particularly when the vomit has a reddish or coffee-colored tinge, indicating the presence of blood. The third is severe diarrhea lasting more than 48 hours, or diarrhea that is bloody or has an unnaturally black, tarry cast. (Blackened stools are a sign of bleeding high up in the digestive tract.) The fourth is fever of 102° F. or more. Any of these symptoms indicates a major infection or organic crisis that may be beyond your ability to treat; it demands fast

A page from a 15th Century Persian medical text includes an illustration of two bristly tamarisk shrubs. The inscription recommends wine made from the berries of the tamarisk as a tonic for treating stomach pain and kidney disease, but urges caution in dosage: Too much wine would produce diarrhea, the text explains, too little would result in constipation.

professional help. And of course anything that does not go away in a week, or that keeps going away and coming back, is something to ask a doctor about.

Assuming that none of these danger signs is present, the best immediate injunction is to do nothing at all. Almost all the digestive problems are what doctors characterize as self-limiting. They pass in a day or two of their own accord without special treatment.

While you are waiting for the system to get itself back in order, there are a number of things you can do on your own to ease discomfort and shorten the wait. For simple indigestion, first of all try to relax. If you feel like lying down, do so. Take no food or drink for a while. If later you feel up to it, try some bland food—broth, dry toast, applesauce and similar dishes (pages 42-43). Avoid butter and other fats. Many people find that tea helps—the principal value seems to come from the hot liquid, so tea should be mild to avoid the irritation its caffeine content can bring.

Informal cures for indigestion abound; many people get help from essence of peppermint and sugar, or even a teaspoon of iced crème de menthe. Licorice is also a time-honored tonic for a nauseated stomach. More certain remedies fill many shelves in the drugstore. But before using them, check their label warnings and consider other medicines you may be taking. Certain drugs themselves cause indigestion, particularly antibiotics and the compounds used to treat arthritis; if this is the source of the problem, your doctor may be able to prescribe a substitute.

Even the medicines meant to cure digestive ailments may introduce complications of their own to the system they are intended to treat. Most of them have to be digested and disposed of after they have completed their work, a waste-treatment process making extra work that can overload and even damage the liver and kidneys (page 38).

Disposal is less a problem with some of the drugs sold to cure simple indigestion, for in a few cases they need no processing by the associated organs; they remain entirely within the digestive tract, going straight through with the food and never entering the tissues of the body itself.

Indigestion remedies are mainly antacids, which ease the stomach by neutralizing its secretions of hydrochloric acid. These neutralizing compounds can soothe the stomach, but they will not cure colds, food intolerances, tension headaches, emotional upsets, morning sickness or other bodily woes as some people believe.

Choosing the right antacid

The ideal antacid works swiftly, lasts a long time, is easy to take, has no side effects and is inexpensive; the fact that so many products vie for attention indicates that none meets all these criteria. Instead, each variety possesses its own particular advantages.

Almost all antacids contain at least one of four basic neutralizing agents, all simple compounds described in high-school chemistry texts—sodium bicarbonate, calcium carbonate, aluminum hydroxide and magnesium hydroxide (also called milk of magnesia). Because these substances work differently and no two people react exactly the same way to them, you may want to try several to find the one that works best with the fewest complications.

One of the oldest and most familiar is sodium bicarbonate—the baking soda found in the supermarket. The major ingredient of two of the top-selling antacids, bicarbonate acts fast and is cheap; the supermarket variety works as well as the costlier drugstore versions. But its effect is fleeting—it lasts about 45 minutes—so that repeated dosage may be required, and this can lead to problems. Unlike some other antacids, sodium bicarbonate is absorbed from the digestive tract into the bloodstream; there it can bring about an imbalance in body chemistry that results in weakness and muscle cramps. Its sodium content can also be harmful to individuals whose heart ailments require salt-free diets.

More potent is calcium carbonate, which is the principal active ingredient in at least one familiar variety of antacid tablet. Longer-lasting than sodium bicarbonate, it is also less absorbed by the blood. An average of only 10 per cent of the calcium enters the system; however, continued use can eventually lead to kidney stones. Calcium carbonate can also be constipating, and although this might appear to recommend it to someone suffering from both stomach upset

The system's extra burden: eliminating medicines

Just as food is chemically changed through digestion before its wastes can be eliminated, so too must most drugs be digested before they can leave the body. In this process, many of the medicines taken to ease digestive troubles put an additional load of their own on an already burdened system, particularly on the liver.

Typical is the processing, illustrated below, that is necessary for the elimination of phenobarbital, a barbiturate sometimes used in combination with other drugs in diarrhea medicines. It circulates through the bloodstream in an insoluble form that cannot be snared by the kidneys. If it and other insoluble drugs are not to pile up into a junkyard of toxic debris, they must be altered in the liver to a form the kidneys can handle.

Liver cells have two types of components, called endoplasmic reticulum: One, rough-surfaced, produces protein; the other, smooth, contains enzymes to transform chemicals into water-soluble forms. The phenobarbital remnant, made soluble by this second component, is carried by the blood to the kidneys, where fluids pass through a sievelike membrane. Filtration tubes allow nutrients and some water to seep back into the bloodstream, leaving water-soluble chemical wastes to be passed out in urine.

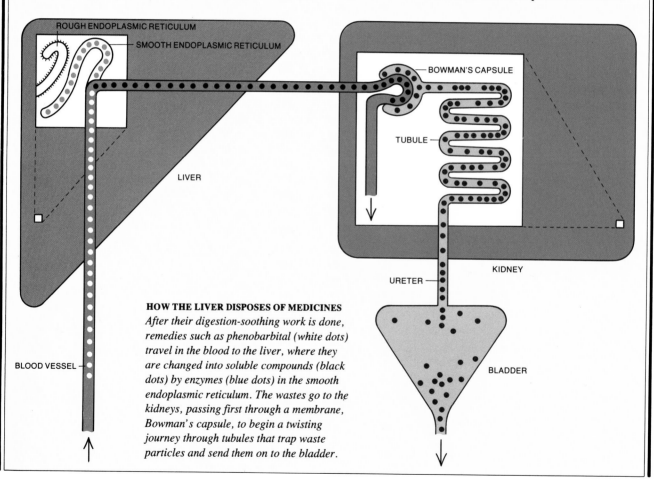

ROUGH ENDOPLASMIC RETICULUM

SMOOTH ENDOPLASMIC RETICULUM

BOWMAN'S CAPSULE

TUBULE

LIVER

KIDNEY

URETER

BLOOD VESSEL

BLADDER

HOW THE LIVER DISPOSES OF MEDICINES
After their digestion-soothing work is done, remedies such as phenobarbital (white dots) travel in the blood to the liver, where they are changed into soluble compounds (black dots) by enzymes (blue dots) in the smooth endoplasmic reticulum. The wastes go to the kidneys, passing first through a membrane, Bowman's capsule, to begin a twisting journey through tubules that trap waste particles and send them on to the bladder.

and diarrhea, most doctors feel that this alone does not warrant its use for this dual purpose. For calcium carbonate is more likely than other antacids to bring the unfortunate effect of acid rebound: Gastric acid, after being temporarily neutralized, presently bounces back to even higher levels as the calcium causes the secretion of gastrin, the hormone that triggers the release of acid.

The magnesium and aluminum compounds that are the other principal antacids are often combined, as they are happily complementary. Both less potent and slower to act than sodium bicarbonate and calcium carbonate, they also raise fewer problems; neither is absorbed into the bloodstream.

Magnesium hydroxide in large doses will generate diarrhea—it is sometimes taken as both a stomach-soother and a laxative. The aluminum compounds, on the other hand, are constipating. Putting the two chemicals together to cancel out each other's drawbacks, a number of manufacturers have marketed joint compounds that are for the most part both safe and effective. One group of medical experts pronounced the magnesium-aluminum products the best.

Extra ingredients that may or may not help

In the hope that extra ingredients will work extra wonders, some manufacturers have incorporated additional chemicals, such as aspirin, into the mix. One popular effervescent product is so made. The theory is that aspirin, the best all-round pain-reliever in the medicine cabinet, will clear up the headache that often accompanies indigestion. There is only one problem: Aspirin irritates the stomach lining and simply adds to the problem the antacid is meant to solve. A government review panel concluded that combination remedies such as the antacid-aspirin drugs were ''irrational for antacid use alone and therefore shall not be labeled or marketed for such use.'' They are certainly not recommended for treating a hangover, for the aspirin intensifies the stomach irritation caused by alcohol. However, some combination antacids contain a different pain-reliever, acetaminophen, which does not bother the stomach.

Another additive is simethicone, which is said to relieve gas pains by reducing the surface tension of air bubbles in the gastrointestinal tract. This action should make small bubbles combine into bigger bubbles, which can be more easily expelled from the body. But while simethicone works fine in the test tube, there is no firm evidence that it eases gas in the gut, where gas tends to concentrate in large pockets anyway. Bloating can best be relieved through diet or behavioral changes rather than through medication. But some doctors believe simethicone has a beneficial placebo effect, prompting the patient to rid himself of excess gas of his own accord because he thinks the medicine is working.

Until recently, caffeine was routinely added to at least one antacid—it will ease certain kinds of headache and may have other analgesic effects. When the government panel of experts pointed out that caffeine increases stomach-acid secretion, the substance was removed. Because of the acid-stimulating effect of caffeine, it makes sense to give up coffee, strong tea and cola drinks—all of which contain caffeine—during stomach upsets. Milk, incidentally, is by no means the ideal stomach-soother it was once considered: Although it has some antacid properties, it leads to acid rebound, stimulating greater acid production in somewhat the same way calcium carbonate does.

Almost as important as the chemical composition of an antacid is the form in which it can be taken. Many products are marketed both as liquids and as chewable tablets, and most others are sold as powders, to be mixed with water before dosing. Liquids and dissolved powders are easier for many people to ingest and usually have a greater neutralizing capacity, and they are therefore better for home use; tablets are handier to carry with you. When using tablets that are meant to be chewed, be sure to chew each one thoroughly for maximum benefit. Also, tablets may have slightly different ingredients from their liquid counterparts; scrutinize labels carefully.

Although stomach sufferers by legend pop their pills while they are eating, the time to take an antacid is not at meals. Food itself will neutralize acid in the stomach for an hour or so, after which an acid rebound will take place. That is when help is needed. The best time to take an antacid is therefore one hour after a meal, followed if necessary by another dose a couple of hours later, plus a final one at bedtime.

If indigestion escalates into true gastritis, antacids and the general treatment for indigestion can help, provided the cause can be identified as a night of heavy drinking, an unwise consumption of overspiced foods or the excessive use of aspirin or other stomach-irritating drugs. The pain should disappear soon enough, since the inflamed stomach lining renews itself every few days.

Other instigators of gastritis, particularly if gastroenteritis is also involved, may be difficult to trace—and to treat. A viral infection of the digestive tract, for example, may bring not only pain and nausea but abdominal cramps, diarrhea, and sometimes fever. There is no cure for a virus. All you can do is go to bed and try to make yourself comfortable by ministering to the symptoms. Here too, antacids may help—but not the magnesium types, which increase intestinal motility, the regular squeezing movements of the intestines that push foods through the system. Drink plenty of liquid: weak tea, broth or gelatin, but not fruit juices, which make the diarrhea worse.

If diarrhea is the principal complaint, remedies in addition to those applied to indigestion may be called for. A multitude

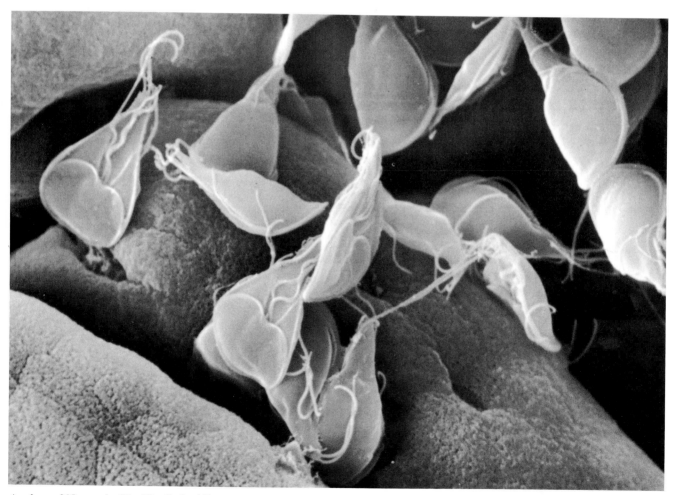

A colony of 15 or so leaflike Giardia lamblia swarm over a section of human intestine. The single-celled parasites cause a sneaky form of traveler's diarrhea: Their incubation period in the intestine is long—up to two weeks—thus many travelers do not come down with symptoms until they have returned home. The drug metronidazole can cure the condition in about a week.

of causes beyond an intestinal virus can bring about diarrhea when excess liquid in the large intestine remains mixed with the waste products of digestion. Nervous tension can speed a meal through the tract before the colon can do its work. Any of the spices or drugs that irritate the stomach can also do damage to the intestinal lining. Some people have a genetic intolerance toward certain foods—notably milk or certain grains—that may incite diarrhea. Meals that are excessively high in fat will sometimes do the same. So may extreme amounts of roughage, such as bran and fibrous vegetables, which are only partly digestible and tend to move quickly through the system; healthful in moderation, they can bring diarrhea when consumed overeagerly. In any of these cases, a change in eating habits will quickly erase the problem.

How to guard against food poisoning

The diarrhea that comes from food poisoning can also bring cramps, vomiting and fever, and the cause is often difficult to identify. The culprit is likely to be one of several types of bacteria that enter the system in spoiled or contaminated food. Some foods are especially prone to spoilage: The so-called scombroid fish—tuna, mackerel, bonito and skipjack—are excellent if eaten fresh, but if these fish are allowed to spoil, their bacteria will produce a toxin that, the American Digestive Disease Society pointed out, "guarantees the eater four or five utterly miserable hours." Beef, pork, poultry and egg dishes, if left unrefrigerated, can develop an organism called Salmonella that raises microscopic ulcers in the intestinal lining. Salmonella causes illness slowly, with an incubation period of from 12 to 48 hours, but it can hit hard. A single tainted chicken sandwich with mayonnaise can lay the victim on his back for several days.

Unpleasant though it is, food poisoning generally clears up of its own accord with no lasting effects. The most dangerous exception is botulism, caused by the toxin produced when common but deadly bacteria multiply in improperly canned food. It is the most potent poison known, but it is destroyed by boiling, and is thus a threat mainly in canned food that is eaten cold. It brings dizziness and a distinctive symptom, double vision. Unless treated immediately it damages the central nervous system and is often fatal. Anyone who suspects he has botulism should go directly to a hospital. The doctor will use cathartics, emetics and enemas to clear out the system as swiftly as possible, and then administer an antitoxin that may halt the attack if used soon enough.

Bivalve shellfish—clams, oysters, mussels and scallops—are the sources of two other kinds of food poisoning with serious effects. Shellfish harvested during the infestations of so-called red tide that occasionally occur in coastal waters contain algae producing a toxin affecting the central nervous system; unlike the poison of botulism, that of red tide is not destroyed by cooking. Most of the time, a well-publicized health alert will warn seafood lovers to avoid these shellfish during red-tide episodes. More insidious is hepatitis, the liver disease that can be contracted by eating raw shellfish from waters contaminated by sewage; the only sure protection against this is personal knowledge of the purity of the shellfish source.

The traveler's bane

One special kind of diarrhea strikes travelers in foreign countries. It is often called *turista* but there are many other names, mostly humorous references to the country where it strikes visitors; in India it is called Delhi belly, in Mexico Montezuma's revenge, and in Russia the Trotzkys. In some cases it is simply bacterial food poisoning, no different from the home-grown variety. But a common villain, particularly in the Caribbean, is either of two single-celled parasites, one a kind of amoeba and the other a type of organism called Giardia; they can work their way into the intestinal lining and cause a lingering type of diarrhea. Both parasites attack by entering the digestive system encapsulated in cysts in contaminated food or water. After about two weeks the organisms emerge from the cysts, like a butterfly from its chrysalis, to begin their noxious work.

Anyone who returns from a foreign visit with a persistent case of diarrhea, either mild or severe, should visit a doctor for tests; if the cysts are present, antibiotics can get rid of the parasites. Because of the frequency with which travelers jet around the world, cases of infection are beginning to spring

Foods that go down easily – and stay down

After a bout of diarrhea or vomiting, the body lacks vital fluids and minerals. In the queasy aftermath, it is important to replace these nutrients by consuming nonirritating foods and liquids. The bland stomach-soothers pictured—all old-time remedies—are still among the best, but as a rule anything the system can handle is suitable. These restoratives—dry white toast, salty crackers, rice, broth, gelatin and selected fruits—go down easily and stay down. Equally important, however, are the water and minerals—including sodium, potassium and chloride compounds—they supply. These crucial chemicals, among the first to go when digestive trouble hits, regulate the body's fluid balance. Folk wisdom ascribes an added attraction to at least one of these foods: Chicken soup, so it is said, is good for almost anything that ails you.

Foods prepared from clear liquids, such as chicken broth and gelatin, relieve the dehydration that often follows diarrhea or vomiting. Dissolved ingredients in these dishes also help get the body back on track: Sodium and chloride compounds in the salty soup keep fluids balanced, and sugar in the gelatin boosts energy.

Salty crackers replenish the body's supply of sodium and chloride. Like the dry white toast at right, the crackers contain complex carbohydrates—a good source of calories for a person weakened by digestive upset. Both types of food are readily digested high in the gastrointestinal tract and thus are unlikely to cause problems farther down.

Because it is digested in the upper gastrointestinal tract, most of this white rice will never reach the colon, so it contributes little additional bulk to worsen diarrhea. The fine, soft grain nourishes the recuperating digestive system, and the whole body, with ample portions of carbohydrates and B vitamins.

Most fruits contain pectin, a chemical that is used in antidiarrhea drugs, but apples—easily digested in sauce form—are especially rich in this desirable ingredient. The banana slices also yield large quantities of potassium, one of the body's fluid-regulators, which is especially depleted by diarrhea.

up in areas where the parasites had been unknown; thus examinations are often given to nontravelers where amoebic dysentery or giardiasis is suspected.

Nearly half of all *turista* problems, however, arise from certain strains of the common bacteria, *Escherichia coli*. Harmless *E. coli* normally frolic in the large intestine with no ill effect—indeed they are necessary tenants, maintaining the balance of bacteria in the gut. The trouble comes when one of the harmful strains, present everywhere but prevalent in the tropics, finds its way into the stomach or small intestine, usually in contaminated drinking water, or in food prepared by unwashed hands. Until the body's natural processes can eradicate the organism, the traveler will find himself running constantly to the bathroom.

The *E. coli* type of *turista* can be prevented with the bismuth salt used to treat diarrhea, which provides a modicum of advance protection, apparently by chemically binding with the bacterial toxin so that it cannot cause harm. A test of United States students traveling to Mexico indicated that imbibing large amounts of the bismuth medicine did in fact keep most of them free of the disease. The hitch was that the students had to drink half a pint of the medicine each day, necessitating a luggage burden that many tourists would prefer to avoid.

Four ways to fight diarrhea

Most cases of diarrhea, including all but the more virulent types of food poisoning, are eminently treatable at home. A variety of medicines is available, but diet is important; most helpful is a liquid diet.

Liquids are needed because the watery waste, far from indicating too much fluid in the system, represents fluids that should have been absorbed into the body by the large intestine. Replace this loss with plain water, clear soups and weak tea. (Be cautious with artificial sweeteners, which may worsen diarrhea.) When solid food appeals, stick to bland foods. Boiled white rice is often recommended, and so are bananas and applesauce. Avoid all other fresh fruits and vegetables.

Among the drugs that help restore normal action to the intestines, the mildest are the so-called adsorbents—substances that combine with the watery feces to thicken them. Among the best known are combinations of kaolin, an aluminum silicate that for generations has been used in the Far East as an antidote for dysentery, and pectin, the substance in citrus fruit and apple pulp that makes jellies jell. Although the kaolin-pectin compounds were found safe by a United States government review panel, their effectiveness for all individuals remains in question. Similar reservations were reported about a slightly stronger group of drugs: the astringents. The most commonly used astringents are bismuth salts, which supposedly solidify protein in the intestines.

If both adsorbents and astringents are found wanting, the best alternative is one of the mild opiates, such as paregoric. A camphorated tincture of opium, paregoric is a mild narcotic; and addicts have been known to get a kick by shooting it directly into their veins. But the government has ruled it safe when taken in the small doses sufficient to counteract diarrhea, and many doctors believe that it and other opiates are among the most effective medicines for this purpose. They seem to work by slowing up intestinal peristaltic activity, allowing time for greater absorption of water. But unless used with discretion, their effect can backfire: If the diarrhea is being caused by infective bacteria, such as Salmonella, the best treatment is to let it move quickly through the system, rather than holding it in the gut and thus prolonging the damage. Because of paregoric's possible misuse, its sale is banned except on a prescription basis, or in extremely mild compounds.

It might be presumed, incidentally, that because some acceptable antacid chemicals, such as aluminum hydroxide, can be constipating they might be used to combat diarrhea, especially if indigestion is also present. The experts who evaluated these drugs for the government recommended against this practice, deciding such use would not be "rational"; just because a drug induces constipation, they said, does not mean it can halt diarrhea. But some gastroenterologists disagree and recommend the medicine for its dual effect.

If nonprescription drugs fail to provide relief within about two days, a physician can order stronger medicine, such as a more potent opiate, or related compounds such as loperamide

Dr. Chevalier Jackson jots notes on a blackboard, long after his first encounters with accidental lye poisoning. Many victims were sent to him because of his unique expertise—he developed a flexible lighted tube for looking inside the esophagus.

A crusader against poisons

Thanks largely to a Philadelphia surgeon, Dr. Chevalier Jackson, officially mandated warnings now protect children against poisonous household chemicals that can sear the digestive tract and even kill *(page 46)*. In the 1890s Dr. Jackson was horrified by the number of children brought to him with their esophagus scarred almost shut by lye, a caustic cleaner then in common use. He devised instruments and surgical techniques to repair their digestive tracts, and sought legislation to help prevent such injuries. After 30 years of lobbying, he was invited to watch as President Coolidge signed the first effective federal poison-labeling law.

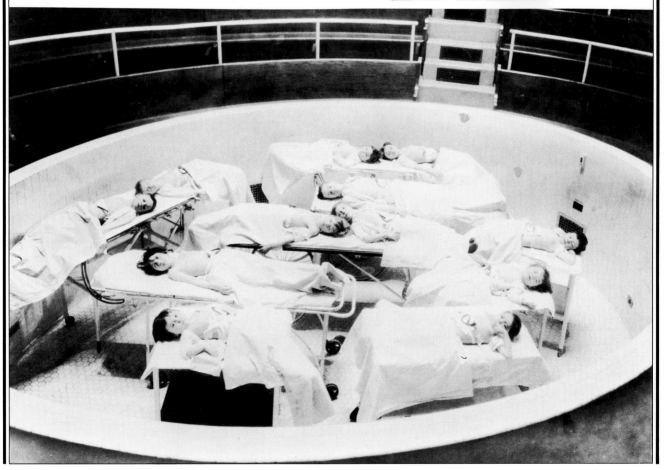

Resting on cots in a surgery amphitheater, Dr. Jackson's young patients recuperate from lye poisoning. Until the burned esophagus healed, each was fed soft foods and liquids through a tube attached to a temporary opening in the stomach.

or diphenoxylate, which also discourage intestinal motility. Or he may try yet another class of drug, the anticholinergics. These agents not only slow down the intestines but also alleviate the painful cramps that often accompany diarrhea and gastroenteritis. The cramps occur when the inflamed intestines go into spasm, their muscular walls contracting violently in response to the trauma occurring within them.

The anticholinergics reduce intestinal spasms by blocking production of a bodily secretion, acetylcholine, that triggers the nerve impulses controlling the intestinal muscles. The most common types are atropine and similar compounds. Like any powerful drug, they have far-reaching side effects: They dry up other important bodily fluids, and can cause drowsiness and blurred vision.

Some mishaps affecting the digestive system have nothing to do with bugs or bad food. The most serious is poisoning by a substance that is not meant to be ingested. Every household harbors many such compounds: ointments in the bathroom cabinet, cleaning and polishing compounds in kitchen and laundry, gasoline and pesticides in the garage. Some are present in the leaves and berries of ornamental plants. Such chemicals constitute a severe risk to young children, who instinctively taste everything they get their hands on; all these substances should not only be kept out of children's reach but also labeled with one of the easily understood signs distributed by poison-control centers *(below)*.

Motion sickness and morning sickness

It is not even necessary to ingest the wrong thing to get an upset stomach. Nausea in particular can represent perfectly normal bodily reactions. One such form is motion sickness. Another is the morning sickness sometimes experienced by pregnant women.

Motion sickness is much less common than it once was, thanks to stabilizers in large ships, smooth-flying jet planes, and modern car suspensions, but it still occurs wherever people are rocked or tumbled about—in automobiles, in small boats, even on roller coasters and other amusement park rides. Most recently, astronauts traveling in space have felt its all-too-familiar dizziness and nausea. The complaint

Fast help for poisoned children

Some of the worst hazards to the digestive system come not from foods but from household poisons swallowed by children. The victims are generally rushed straight to the hospital. Yet they might get swifter care if a parent first called the nearest poison-control center (listed in most telephone books with emergency numbers). Such centers are clearinghouses for data on the dangerous chemicals that lurk in every home.

The experts who run the centers recommend that parents keep handy a bottle of syrup of ipecac—a vomit-inducing agent available at pharmacies—but use it only when advised to by the poison-control specialist who, if told what substance was swallowed, will know whether vomiting will help.

Printed with hot-line numbers, graphic stickers warning children away from poisons are available at many control centers. The face of "Mr. Yuk" (left) or "Officer Ugg" (right) delivers a more understandable hands-off message than official printed warnings or the traditional skull and bones, which many toddlers associate with amusement parks and movie pirates.

arises when the semicircular canals in the vestibule of the ear, which control balance, are thrown off kilter by motion to which the body is not accustomed. The disturbance is communicated to the brain center that controls vomiting, and queasiness soon follows.

Rules for preventing motion sickness abound. The most sensible of these command you to eat lightly before embarking, but eat something, since an empty stomach succumbs more easily; stay in the open if possible, away from kitchen odors and engine fumes; and fix your eyes on the horizon, a remarkably stabilizing sight. Above all, keep busy. Said one naval officer, "In my opinion, chipping paint from a bulkhead is the best cure for seasickness in young sailors."

Several nonprescription medicines will prevent, or at least reduce, the unpleasantness. Most contain one of the three compounds that government experts found best suited to the task: cyclizine, meclizine or dimenhydrinate. All three are believed to work by suppressing the sensitivity of the balance mechanism in the ear. They enjoy a reputation for successfully accomplishing their purpose if taken before the journey starts.

Morning sickness, the second benign distress, afflicts about half of all expectant mothers. No one knows exactly why some pregnant women should feel ill—or why the other half breeze through pregnancy untroubled. Certainly the system undergoes chemical changes during gestation. Some doctors call the nausea psychosomatic, and there is general agreement that expectant mothers who are nervous and unhappy women are more susceptible to the ailment. In most women, morning sickness is little more than an annoyance: Those who get it feel nauseated when they awaken, but as soon as they have breakfast the feeling disappears. The affliction does not interfere with nutrition and the vomiting it sometimes involves cannot, as some young mothers fear, cause a spontaneous abortion.

Because an empty stomach contributes to the problem, doctors have a simple remedy. They advise their patients to keep a box of crackers at bedside and nibble a few on awakening, then stay in bed for 20 minutes to allow the food to start digesting. By that time the queasiness should be gone.

Most women find the problem disappears after the first three months of pregnancy. Some doctors even look with favor on morning sickness, convinced that it promises an easier-than-average delivery.

Irregularity: an overrated problem

The various forms of nausea, like nearly all the complaints that plague the digestive system, are a sign of inappropriate activity somewhere along the line. But the complaint that causes the most—and, generally, the most undue—perplexity is a decrease in activity: constipation. Nonprescription laxatives sell at a rate of around $400 million worth a year in the United States. Most doctors believe that this expense is for the most part needless, if not harmful. It is true that some constipation can be organic, resulting from an intestinal tumor or abnormal operation of the thyroid gland, which controls the body's use of food. But these complications are rare compared to the constipation that is self-induced—or nonexistent except in the imagination. Furthermore, it could generally be prevented by simple measures.

The unnecessary concern over constipation arises from the widespread but erroneous notion that a daily bowel movement, performed with clockwork regularity, is essential to good health. The idea is probably as old as civilization itself; the ancient Egyptians, believing that food was the source of all illness, subjected themselves to a thrice-monthly purge of castor oil mixed with beer. Today's laxative advertisers, noted the U.S. Food and Drug Administration, have contributed to the problem by promoting super-regularity.

But individual rates of motility differ widely, and a perfectly normal rate of bowel movement can vary anywhere from three times a day to three times a week. One physician recalled a dinner party interrupted by an "emergency" phone call from a patient who said she could not go to sleep because she had not had a bowel movement for 36 hours. "I was very much tempted to tell her I hadn't either," said the doctor. "I sometimes think people would be happier and healthier if they didn't know constipation existed."

Certainly history's most famous constipation victim, Martin Luther, suffered woefully and often wrote about his complaint. ''The Lord thus afflicts me,'' he declared, ''that I may not be without a relic of the cross. May he be blessed.''

The mistaken veneration for regularity, then as now, is based on the belief that intestinal waste, if allowed to remain in the colon for more than a day or so, can poison the system, a process referred to as auto-intoxication. The assumption is false. Although the large intestine's contents are alive with bacteria, these not only are harmless while still in the tract but are essential, and it makes no difference how long they remain. People who have gone camping in the Arctic have voluntarily abstained from moving their bowels in the frigid outdoors for as long as three weeks with no ill effects.

In an attempt to explore—and perhaps explode—the regularity myth, Dr. Walter C. Alvarez of the University of Minnesota once conducted an interesting experiment with a group of his students, all in good health. They swallowed gelatin capsules containing small glass beads. Only two of the volunteer subjects passed nearly all their beads in the first 24 hours. Most took four days to eliminate as many as three fourths of them. Some students, in fact, had after nine days eliminated only half their beads.

Any number of factors, from changes in daily routine to dietary variations, can affect the rhythm. One way to think of the intestinal tract, Dr. Alvarez commented, is to liken it to a railroad siding on which a number of boxcars are standing. When a new car arrives it will bump the end car off, leaving the same number of cars on the siding. Some days no new car will arrive. But sometimes a car will bump the others with such force—as when a large meal full of fiber or other roughage induces a large bowel movement—that most of the cars are knocked off. Laxatives have this effect. Several days may then pass before another boxcar is ready to be dislodged. Viewed in this way, irregularity is an acceptable norm.

Certain substances or events can slow intestinal motility. Lack of exercise, for example, can cause constipation. Various drugs, including codeine and several other prescription painkillers, a few antacids, tranquilizers and certain medicines for high blood pressure, have the same effect. Psychological turmoil can contribute to constipation as well as to diarrhea, and can even cause the two complaints to alternate;

this condition is called irritable bowel syndrome and is most readily treated by modifying diet and behavior (Chapter 3). Traveling or changes in daily habits affect motility for a while. And older people generally experience a slowing down of their digestive processes, in some cases because they suffer from other ailments but more likely from an unbalanced diet or from lack of exercise.

Too often, however, constipation is provoked by the individual sufferer because he disregards body signals and disrupts natural rhythms. A more serious cause is laxative abuse. An astonishing number of people resort to laxatives habitually, often because they dread the imagined consequences of irregularity. In the process they not only build up a dependence on the medication—their colons lose the ability to respond unless stimulated—but risk damage to their intestines.

The harm that excessive use of laxatives can bring is illustrated by the notorious case of a 42-year-old woman whose colon seemed, from X-rays and other tests, to exhibit characteristics of cancer and of chronic inflammation. Doctors removed her colon. Study of the excised tissue revealed that she had no malignancy. Earlier she had admitted to taking laxatives for four years. The doctors, probing further, questioned her husband closely and learned that she had actually been taking a potent stimulant laxative for no less than 18 years. She had visited upon herself a cathartic colon—one battered by laxative abuse.

A laxative may occasionally be needed, but the best way to avoid constipation is with proper living habits: Eat enough roughage—vegetables, fruit and whole-grain breads and cereals—to assure good motility; drink plenty of fluids; get plenty of exercise.

Laxatives and cathartics

If a laxative is called for, use the mildest variety available. Among the least damaging are the so-called bulk-formers. These fibrous substances soak up water in the intestines and expand to create jelly-like compounds that increase the size and softness of fecal matter. Bran acts this way. Some bulk producers sold over the counter are cellulose derivatives.

Others are made from psyllium seeds, taken from various of the semitropical herbs called Plantago and from other fibrous plants.

Equally benign are the stool-softeners. Most of them, composed of a chemical called docusate, allow fluids to penetrate the feces, softening them for easier elimination. Old-fashioned mineral oil, which acts in a similar way, is now rarely recommended, as it can interfere with the absorption of fat-soluble vitamins.

Quicker to act are the saline laxatives, or salts: various phosphate and magnesium compounds. Salts stimulate secretion of water into the digestive tract, thus quickening peristaltic activity. Many physicians feel that saline laxatives should not be taken except on a doctor's orders.

The strongest, and potentially the most dangerous, laxatives are the stimulant cathartics. They achieve their effect either by irritating the surface of the intestinal wall or by stimulating nerves along the tract—doctors are not sure which. The result is a great increase in motility, but a frequent side effect is painful cramping.

Among the stimulants are phenolphthalein and the anthraquinones, which include senna and cascara. One traditional stimulant, castor oil, is no longer recommended for simple constipation because it can deplete the body of vital fluids and nutrients.

The stimulant cathartics have been called the most abused of all the laxatives. One woman who visited an orthopedic clinic complaining of back and hip pain was diagnosed as suffering from osteomalacia, a softening of the bones. It turned out that she had been taking 15 to 20 phenolphthalein tablets a day for 20 years, impairing absorption of the vitamins and minerals that bones need. When she discontinued the medicine, her bowel habits returned to normal and her back and hip trouble disappeared.

This case is certainly an extreme. But it illustrates the hazard of imposing rigid, artificial control on a system that is not supposed to operate exactly the same way two days in a row. The proper treatment for an occasional digestive complaint is simply a respite for the tract. Let the body function naturally—which is, of course, what it does best. ✳

A medicine-filled wine glass is raised as if in a toast in this 19th Century poster advertising Dr. Roback's stomach remedy. So much irritating alcohol was in such bitters—some were 88 proof—that the potions could aggravate indigestion while drugging users to forget their pangs. Modern medicines rarely offer such an alcoholic high, but they work better (overleaf).

Drugs that ease digestive woes

Drugstore shelves are loaded with medicines for the digestive tract. Doctors can prescribe scores more. The table below, prepared by Christopher S. Conner of the Rocky Mountain Drug Consultation Center, lists those drugs most widely used for disorders of the digestive tract. They are identified by their generic chemical names; if the drug requires a prescription, it appears with the pharmaceutical insignia, Rx.

Almost all of these remedies can be effective. Perhaps because of that, many are overused. In the quest for regularity, for example, many people take laxatives when they are not needed. Other people, fearing the discomfort of indigestion, pop antacids after every meal. Such practices are dangerous—prolonged use of many of these products can induce psychological or physical dependence on them, or even bring harm to vital organs.

All drugs have side effects. Use the third and fourth columns of the table to check on a drug's minor and major side effects—that is, those conditions that might arise with normal use of the drug in a healthy person. The special-cautions column contains specific warnings on use, particularly for those people with pre-existing conditions—such as heart, lung or kidney trouble—that are unrelated to the ailment for which the drug is intended.

When using any drug, inform a physician of symptoms that worsen or do not improve within a few days. As a general rule, do not use over-the-counter digestive drugs for more than one week continuously. And remember these basic cautions: Pregnant women and nursing mothers should consult a doctor before taking any drug. Drugs that cause drowsiness should not be mixed with alcohol, nor should anyone taking such a drug drive or operate heavy machinery. Consult a physician if serious side effects appear, or if you experience unusual or inexplicable symptoms.

DRUG	Intended effect	Minor side effects	Serious side effects	Special cautions
ALUMINUM HYDROXIDE **AMPHOJEL** **DI-GEL*** **GELUSIL*** **MAALOX*** **MYLANTA***	Neutralizes stomach acid	Constipation; nausea; vomiting	Severe constipation; fecal impaction; bone disease with prolonged use	Consult doctor before taking if you have kidney disease. Take within one hour after meals to maximize duration of effects.
ALUMINUM PHOSPHATE **PHOSPHALJEL**	All effects similar to ALUMINUM HYDROXIDE			
AMOXICILLIN (Rx) **AMOXIL**	Cures urinary-tract infections	Nausea; diarrhea; vomiting	Allergic reactions, such as skin rash, itching, wheezing; inflammation of the colon (colitis)	Consult doctor before taking if you have any allergies, particularly to penicillin, or if you have kidney disease. Keep taking this medicine until it is gone, even if you feel better in a few days. Notify doctor if you are troubled by persistent or severe diarrhea—a sign of colitis.
AMPICILLIN (Rx) **AMCILL** **POLYCILLIN** **PRINCIPEN**	All effects similar to AMOXICILLIN			
ATROPINE (Rx) **BARBIDONNA*** **DONNATAL***	Reduces irritability and spasm of digestive tract; controls incontinence	Constipation; flushed skin; drowsiness; dry mouth; blurred vision; nervousness; reduced perspiration	Persistent difficulty urinating; confusion and delirium; hallucinations; increased pressure inside the eyeball (glaucoma); irregular heartbeat	Consult doctor before taking if you have heart disease, ulcer bronchitis, glaucoma, colitis or urinary or intestinal obstruction. Discontinue use and notify your doctor if you develop eye pain, confusion or difficulty urinating. Reduced perspiration can make exercising or working in hot weather dangerous.

* Combination drug. Refer also to other active ingredients on label.

DRUG	Intended effect	Minor side effects	Serious side effects	Special cautions
BENZOCAINE AMERICAINE HEMORRHOIDAL*	Relieves pain and itching of hemorrhoids	Local burning, tenderness or irritation	Allergic reactions, such as skin rash, itching, wheezing, hives, difficulty breathing; nervousness, dizziness or blurred vision with high doses	Consult doctor before taking if you have allergies to local anesthetics, or if you suffer from asthma.
BETHANECHOL (Rx) DUVOID URECHOLINE	Relieves difficult urination	Diarrhea; headaches; flushed skin; excessive salivation and sweating; blurred vision	Decreased blood pressure; shortness of breath; tightness in chest	Consult doctor before taking if you have asthma, low blood pressure, heart disease, thyroid disease, ulcer or obstruction of the urinary tract. Inform doctor of lightheadedness or dizziness—signs of low blood pressure.
BISACODYL DULCOLAX	Relieves constipation	Diarrhea; nausea	Colic and severe diarrhea with prolonged use; dehydration; bone disease; inflammation of the colon (colitis); tiredness or weakness; irregular heartbeat	Consult doctor before taking if you have heart disease, rectal bleeding or intestinal obstruction. Swallow tablets whole (do not chew) to avoid stomach irritation. Take at least 1 hour apart from antacids or milk.
BISMUTH SUBSALICYLATE PEPTO-BISMOL	Relieves diarrhea; used to prevent traveler's diarrhea	Stool discoloration	Impaction of feces, mainly in infants and the elderly; ringing in the ears, rapid breathing, dizziness or confusion—signs of toxicity	Consult doctor before taking if you have intestinal obstruction, colitis, or are taking aspirin. Do not use for longer than two days without consulting your doctor.
BUTABARBITAL (Rx) BUTIBEL* CYSTOSPAZ-SR* SIDONNA*	Relieves anxiety and, in combination with other ingredients, helps reduce irritability and spasm of digestive tract	Dizziness; clumsiness; drowsiness	Difficulty breathing; confusion; excitement; liver damage; reduced blood-cell counts	Consult doctor before taking if you have lung or liver disease. Dependence may occur with prolonged use at high doses. Do not discontinue abruptly after extended use at high doses—withdrawal symptoms, such as agitation and seizures, may occur.
CALCIUM CARBONATE TITRALAC TUMS	Neutralizes stomach acid	Flatulence; constipation; nausea; vomiting; rebound effect—renewed acidity that may be worse than the original condition	Impaction of feces with prolonged use; severe nausea, vomiting and kidney damage if taken for a prolonged period in conjunction with other antacids, such as sodium bicarbonate, or with milk	Consult doctor before taking if you have kidney disease.
CALCIUM POLYCARBOPHIL MITROLAN	Relieves diarrhea and constipation	Abdominal fullness; nausea; vomiting	None	Consult doctor before taking if you have rectal bleeding or intestinal obstruction.
CARBENICILLIN INDANYL SODIUM(Rx) GEOCILLIN	Cures urinary-tract infections and inflammation of the prostate gland	Nausea; vomiting; diarrhea; bitter taste in mouth	Allergic reactions, such as skin rash, hives, itching and wheezing	Consult doctor before taking if you have any allergies, especially to penicillin. Keep taking this drug as directed until it is gone, even if you feel better in a few days.

52

DRUG	Intended effect	Minor side effects	Serious side effects	Special cautions
CEPHALEXIN (Rx) **KEFLEX**	Cures urinary-tract infections	Diarrhea; nausea; vomiting	Allergic reactions, such as skin rash, hives, itching and wheezing	Consult doctor before taking if you have any allergies, particularly to penicillin. Take with food if upset stomach occurs. Keep taking this medicine as directed until it is gone, even if you feel better in a few days.
CHLORDIAZEPOXIDE (Rx) **LIBRAX***	Relieves anxiety and tension accompanying gastrointestinal disturbances	Drowsiness; slurred speech; weakness; clumsiness	Depression; confusion; excited or agitated behavior; difficulty breathing; liver damage; reduced white-blood-cell count	Prolonged use of this drug can cause psychological and physical addiction even at recommended doses. Consult doctor before discontinuing this drug after extended use—withdrawal symptoms can occur, including agitation, confusion and seizures.
CHOLESTYRAMINE (Rx) **QUESTRAN**	Relieves diarrhea	Constipation; belching; indigestion; nausea; vomiting	Severe constipation or fecal impaction; stomach bleeding; gallstones; pancreas inflammation; weight loss	Consult doctor before taking if you suffer from any of the following: gallstones, pancreatic disease, gastrointestinal disorders or kidney disease. Must be mixed with water or other fluids before taking; swallowing dry powder may cause blockage of the esophagus. Inform doctor of black, tarry stools—signs of stomach bleeding.
CIMETIDINE (Rx) **TAGAMET**	Controls ulcers	Diarrhea; dizziness; headache	Behavior disturbances, including confusion and delirium; kidney and liver damage; reduced white-blood-cell count; breast enlargement or soreness	Consult doctor before taking if you have liver or kidney disease, or if you are 60 years of age or older. Take this drug with or just after meals. Do not take antacids within 1 to 2 hours of taking cimetidine.
CINOXACIN (Rx) **CINOBAC**	Cures urinary-tract infections	Nausea; dizziness; diarrhea	Liver disease; vision changes; reduced white-blood-cell and platelet counts	Take with a full glass of water and on an empty stomach. Keep taking this medicine as directed until it is gone, even if you feel better in a few days. Dizziness may make driving or operating heavy machinery dangerous. Inform doctor of sore throat, bleeding or bruising—signs of reduced blood-cell counts.
CLIDINIUM BROMIDE (Rx) **LIBRAX***	Relieves spasms and discomfort in digestive tract; relieves ulcers; controls incontinence	Dizziness; dry mouth; drowsiness; blurred vision	Persistent constipation; difficulty urinating; severe skin rash; increased pressure inside the eye (glaucoma)	Consult doctor before taking if you have colitis, intestinal obstruction, glaucoma or urinary obstruction. Take 30 minutes before meals unless otherwise directed.
DANTHRON **DORBANE** **MODANE**	Relieves constipation	Nausea; diarrhea; harmless urine discoloration, usually pinkish or red	Tiredness; weakness; irregular heartbeat; severe diarrhea with prolonged use; muscle cramps	Consult doctor before taking if you are breast feeding or have heart disease, rectal bleeding or intestinal obstruction. Take this drug on an empty stomach.
DEHYDROCHOLIC ACID **DECHOLIN** **NEOCHOLAN**	Relieves constipation; helps digestion and absorption of nutrients	Diarrhea	Allergic reactions, such as itching, skin rash or wheezing	Consult doctor before taking if you have liver disease; obstruction of biliary, intestinal or urinary tract; or allergy to tartrazine—a dye used in drugs and foods. Prolonged use may result in dependence. Laxative effect may not occur for several days.

* Combination drug. Refer also to other active ingredients on label.

DRUG	Intended effect	Minor side effects	Serious side effects	Special cautions
DEXAMETHASONE (Rx) DECADRON DEXONE HEXADROL	Relieves inflammation of the colon (colitis)	Nausea; indigestion; insomnia; weight gain; muscle cramps; menstrual irregularities	Depression or emotional disturbances; potassium loss; acne; elevated blood pressure; ulcer; bone disease; pancreas inflammation; increased pressure in the eye (glaucoma); impaired immune response; cataracts	Consult your physician before taking this drug if you suffer from glaucoma. Adhere to a low-salt diet. Inform doctor of black, tarry stools or persistent stomach pains—signs of bleeding from the stomach or intestine. Do not discontinue abruptly after prolonged use—this can cause adverse reactions, such as fever, weakness and dangerous decreases in blood pressure. Do not submit to any vaccinations or skin tests without consulting your doctor.
DIHYDROXYALU-MINUM AMINOACETATE ROBALATE	All effects similar to ALUMINUM HYDROXIDE			
DIHYDROXYALU-MINUM SODIUM CARBONATE ROLAIDS	All effects similar to ALUMINUM HYDROXIDE and SODIUM BICARBONATE			
DIMENHYDRINATE DRAMAMINE	Controls vomiting and prevents motion sickness	Drowsiness; dizziness; dry mouth; nervousness or insomnia, particularly in children	Irregular heartbeat; hallucinations; confusion; delirium	Consult doctor before taking if you have glaucoma. Take at least 1 hour before travel for maximum effect.
DIPHENHYDRAMINE (Rx) BENADRYL	Controls nausea and vomiting	Drowsiness; dizziness; dry mouth; difficulty urinating	Irregular heartbeat; hallucinations; confusion; delirium	Consult doctor before taking if you have glaucoma, high blood pressure, heart disease or urinary obstruction.
DIPHENOXYLATE (Rx) COLONIL* LOMOTIL*	Controls diarrhea and abdominal cramps	Drowsiness; constipation; dizziness; headache	Intestinal obstruction	Consult doctor before taking if you have liver disease or inflammation of the colon (colitis). This drug can cause psychological and physical dependence with extended use at high doses. Exercise extreme caution in using this drug for infants and children—they are highly susceptible to side effects.
DOCUSATE CALCIUM DOCUSATE POTAS-SIUM DOCUSATE SODIUM COLACE KASOF SURFAK	Relieves constipation	Diarrhea	None	This drug may require 24 to 48 hours for maximum effect. Mineral oil should not be used with this drug—absorption of mineral oil may be increased, harming the liver.
DOXYCYCLINE (Rx) VIBRAMYCIN	Used to prevent traveler's diarrhea	Nausea; diarrhea; stomach cramps; heartburn; increased sensitivity to sunlight	Inflammation of the colon (colitis); allergic reactions such as skin rash, hives, wheezing; reduced red- and white-blood-cell counts; severe headache; blurred vision; kidney damage	Consult doctor before taking if you have any allergies. Avoid use for children under 8 years of age—it can discolor their teeth. Discard unused preparations that are older than 3 months—they can produce kidney damage.

DRUG	Intended effect	Minor side effects	Serious side effects	Special cautions
GLYCOPYRROLATE (Rx) **ROBINUL**	All effects similar to CLIDINIUM BROMIDE			
HYDROXYZINE (Rx) **ATARAX** **VISTARIL**	Relieves nausea and vomiting	Drowsiness; dry mouth	Skin rash; muscle tremors	Drowsiness can make driving or operating heavy machinery dangerous.
HYOSCYAMINE (Rx) **CYSTOSPAZ** **LEVSIN**	All effects similar to ATROPINE			
ISOPROPAMIDE (Rx) **COMBID*** **DARBID***	All effects similar to CLIDINIUM BROMIDE			
KAOLIN-PECTIN **DONNAGEL*** **DONNAGEL-PG* (Rx)** **KAOPECTATE** **PARGEL**	Relieves diarrhea	Constipation	Severe constipation and fecal impaction, mainly in infants and the elderly	Consult doctor before taking if you have stomach ulcer or heart disease. Inform physician if diarrhea is not relieved within 48 hours.
KETOCHOLANIC ACID (Rx) **KETOCHOL**	All effects similar to DEHYDROCHOLIC ACID			
LACTULOSE (Rx) **CHRONULAC**	Relieves constipation	Flatulence; abdominal cramps; nausea; increased thirst	Severe diarrhea and dehydration; tiredness; weakness	Consult doctor before taking if you suffer from any of the following ailments: heart disease, rectal bleeding, intestinal obstruction or diabetes mellitus. This drug may take 1 to 2 days to produce maximum effects. Mix with fruit juice, milk or water to improve flavor. Inform doctor of persistent diarrhea.
LOPERAMIDE (Rx) **IMODIUM**	Controls diarrhea	Nausea; vomiting; drowsiness; dry mouth	Severe constipation; appetite loss	Consult doctor before taking if you have liver disease, inflammation of the colon (colitis) or diarrhea due to bacterial organisms. Notify doctor if diarrhea persists or if fever develops while you are taking this medicine.
MAGNESIUM HYDROXIDE **DELCID*** **DI-GEL*** **GELUSIL*** **MILK OF MAGNESIA** **MYLANTA***	Neutralizes stomach acid; (Milk of Magnesia also used as a laxative)	Diarrhea	Decreased blood pressure, drowsiness, nausea and vomiting—usually in those with kidney disease	Consult doctor before taking if you have kidney disease.
MAGNESIUM TRISILICATE **TRISOGEL***	All effects similar to MAGNESIUM HYDROXIDE			
METHANTHELINE (Rx) **BANTHINE**	All effects similar to CLIDINIUM BROMIDE			

* Combination drug. Refer also to other active ingredients on label.

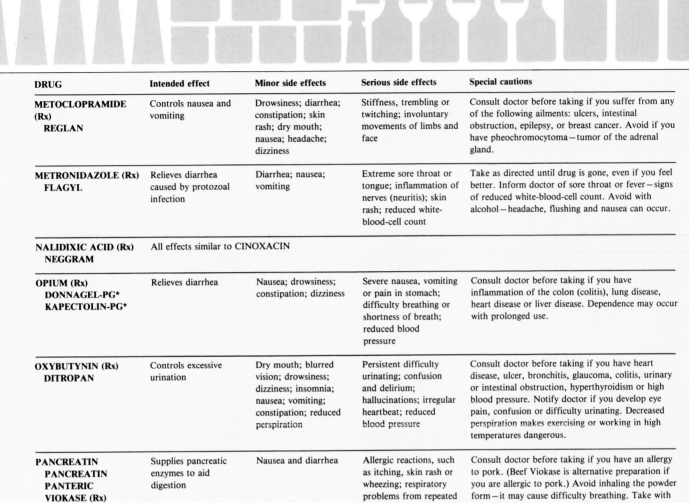

DRUG	Intended effect	Minor side effects	Serious side effects	Special cautions
METOCLOPRAMIDE (Rx) REGLAN	Controls nausea and vomiting	Drowsiness; diarrhea; constipation; skin rash; dry mouth; nausea; headache; dizziness	Stiffness, trembling or twitching; involuntary movements of limbs and face	Consult doctor before taking if you suffer from any of the following ailments: ulcers, intestinal obstruction, epilepsy, or breast cancer. Avoid if you have pheochromocytoma—tumor of the adrenal gland.
METRONIDAZOLE (Rx) FLAGYL	Relieves diarrhea caused by protozoal infection	Diarrhea; nausea; vomiting	Extreme sore throat or tongue; inflammation of nerves (neuritis); skin rash; reduced white-blood-cell count	Take as directed until drug is gone, even if you feel better. Inform doctor of sore throat or fever—signs of reduced white-blood-cell count. Avoid with alcohol—headache, flushing and nausea can occur.
NALIDIXIC ACID (Rx) NEGGRAM	All effects similar to CINOXACIN			
OPIUM (Rx) DONNAGEL-PG* KAPECTOLIN-PG*	Relieves diarrhea	Nausea; drowsiness; constipation; dizziness	Severe nausea, vomiting or pain in stomach; difficulty breathing or shortness of breath; reduced blood pressure	Consult doctor before taking if you have inflammation of the colon (colitis), lung disease, heart disease or liver disease. Dependence may occur with prolonged use.
OXYBUTYNIN (Rx) DITROPAN	Controls excessive urination	Dry mouth; blurred vision; drowsiness; dizziness; insomnia; nausea; vomiting; constipation; reduced perspiration	Persistent difficulty urinating; confusion and delirium; hallucinations; irregular heartbeat; reduced blood pressure	Consult doctor before taking if you have heart disease, ulcer, bronchitis, glaucoma, colitis, urinary or intestinal obstruction, hyperthyroidism or high blood pressure. Notify doctor if you develop eye pain, confusion or difficulty urinating. Decreased perspiration makes exercising or working in high temperatures dangerous.
PANCREATIN PANCREATIN PANTERIC VIOKASE (Rx) BEEF VIOKASE (Rx)	Supplies pancreatic enzymes to aid digestion	Nausea and diarrhea	Allergic reactions, such as itching, skin rash or wheezing; respiratory problems from repeated inhalation of the powder	Consult doctor before taking if you have an allergy to pork. (Beef Viokase is alternative preparation if you are allergic to pork.) Avoid inhaling the powder form—it may cause difficulty breathing. Take with or just before meals.
PANCRELIPASE (Rx) COTAZYM ILOZYME PANCREASE	All effects similar to PANCREATIN			
PAREGORIC (Rx) PAREPECTOLIN*	All effects similar to OPIUM			
PHENAZOPYRIDINE (Rx) PYRIDIUM	Relieves discomfort and pain of urinary-tract irritation	Stomach cramps; dizziness; headache; urine discoloration, which is harmless	Anemia; kidney disease; liver disease	Consult doctor if you develop yellowing of the skin or eyes—a sign of liver damage. Take with meals to lessen stomach upset.
PHENOBARBITAL (Rx) BELLADENAL* BELLERGAL* DONNATAL*	All effects similar to BUTABARBITAL			

DRUG	Intended effect	Minor side effects	Serious side effects	Special cautions
PHENOLPHTHALEIN ALOPHEN CORRECTOL* EX-LAX	Relieves constipation	Nausea; diarrhea; harmless urine discoloration; abdominal cramping	Allergic reactions, such as skin rash and itching; severe diarrhea, colic, dehydration or potassium loss with prolonged use	Laxative effect may last up to 3 days. Take on an empty stomach for faster results.
PRAMOXINE ANUSOL*	All effects similar to BENZOCAINE			
PREDNISOLONE (Rx) DELTA-CORTE HYDELTRA STERANE	All effects similar to DEXAMETHASONE			
PREDNISONE (Rx) DELTASONE ORASONE PARACORT	All effects similar to DEXAMETHASONE			
PROCHLORPERAZINE (Rx) COMBID* COMPAZINE	Relieves nausea and vomiting	Drowsiness; dry mouth; nasal congestion; difficulty urinating; decreased perspiration; constipation	Decreased white-blood-cell count; liver disease (hepatitis); blurred vision; stiffness; trembling; muscle spasms or twitching; restlessness; decreased blood pressure; skin reactions; severe dizziness	Consult doctor before taking if you have blood disorders, liver disease, glaucoma, epilepsy, heart disease or lung disease. Consult doctor if you develop sore throat or fever—signs of low white-blood-cell count. Consult doctor if you feel faint, experience heart palpitations, or become dizzy. Avoid excessive exposure to heat or exercising in hot weather. Consult doctor if you develop yellowing of the whites of the eyes or skin—signs of liver damage.
PROPANTHELINE (Rx) PRO-BANTHINE	All effects similar to CLIDINIUM BROMIDE			
PSYLLIUM EFFERSYLLIUM METAMUCIL MODANE BULK	Relieves constipation	Abdominal cramps; nausea; diarrhea	Blockage of esophagus or intestines	Consult doctor before taking if you have any of the following ailments: difficulty swallowing (dysphagia), rectal bleeding, heart disease or intestinal obstruction. Take with a full glass of water, and drink 6 to 8 full glasses of water daily. Laxative effect usually occurs in 12 to 24 hours, but may take up to 3 days.
SCOPOLAMINE (Rx) BARBIDONNA* DONNATAL*	All effects similar to ATROPINE			
SENNA SENOKOT	Relieves constipation	Nausea; abdominal cramps; harmless urine discoloration	Tiredness, weakness, muscle cramps or mental confusion with prolonged use	Consult doctor before taking if you have any of the following ailments: heart disease, high blood pressure or intestinal obstruction. Take with a full glass of water. Drink 6 to 8 glasses of water a day.
SIMETHICONE MYLICON	Relieves gas	Belching; flatulence	None	None

* Combination drug. Refer also to other active ingredients on label.

DRUG	Intended effect	Minor side effects	Serious side effects	Special cautions
SODIUM BICARBONATE **ALKA-SELTZER*** **ALKA-SELTZER WITHOUT ASA*** **CITROCARBONATE*** **SODA MINT**	Neutralizes stomach acid	Flatulence; rebound effect—renewed acid that may be worse than the original condition	Fluid retention and weight gain; swelling in feet and legs; blood-chemistry changes; headache, dizziness, confusion, tiredness or convulsions—signs of too much sodium; nausea, vomiting and kidney damage due to prolonged use with calcium carbonate or milk	Consult doctor before taking if you have kidney or heart disease, or if you are on a low-sodium diet.
SUCRALFATE (Rx) **CARAFATE**	Controls duodenal ulcers	Constipation	None	Take at least ½ hour apart from antacids.
SULFASALAZINE (Rx) **AZULFIDINE** **SAS-500**	Relieves inflammation of the colon (colitis)	Nausea; vomiting; headache; loss of appetite	Reduced blood-cell counts; severe skin reactions; fever; liver damage; kidney damage	Consult doctor before taking if you have liver or kidney disease, blood disorders, or intestinal- or urinary-tract obstructions. Drink 6 to 8 glasses of water a day while taking this medicine. Take with food to reduce stomach upset. Inform doctor of weakness, sore throat, fever or unusual bruising or bleeding—signs of reduced blood-cell counts. Inform doctor of yellowing of the whites of the eyes or skin—signs of liver damage.
SULFISOXAZOLE (Rx) **GANTRISIN**	Cures urinary-tract infections	Nausea; vomiting; diarrhea	Allergic reactions, such as skin rash, hives, wheezing; muscle aches and pains; reduced blood-cell counts; severe skin reactions; liver or kidney damage	Drink plenty of fluids—6 glasses of water per day—while taking this medicine. Keep taking this medicine as directed until it is gone, even if you feel better in a few days. Consult doctor if you develop weakness, sore throat, or unusual bleeding or bruising—signs of reduced blood-cell counts.
TRIMETHOBENZA-MIDE (Rx) **TIGAN**	Relieves nausea and vomiting	Drowsiness; dizziness; headache	Depression; liver damage; severe trembling and stiffness of arms and legs; allergic reactions, such as skin rash	Notify doctor of yellowing of the whites of the eyes or skin—signs of liver damage.
TRIMETHOPRIM and SULFAMETHOXA-ZOLE (Rx) **BACTRIM** **SEPTRA**	Cures urinary-tract and kidney infections	Nausea; vomiting; diarrhea	Allergic reactions such as skin rash, hives, wheezing; severe muscle aches and pains; reduced blood-cell counts; liver disease; disturbed kidney function	Drink at least 3 glasses of water a day while taking this medicine. Keep taking this medicine until it is gone, even if you feel better in a few days. Notify doctor if you develop weakness, sore throat, or unusual bleeding or bruising—signs of reduced blood-cell counts.
YEAST CELL DERIVATIVE **PREPARATION H***	Relieves pain and itching of hemorrhoids	Skin rash	None	None

CHAPTER 3

Calming a tract under stress

From pressure and anxiety: queasiness
Why stomach acid goes astray
Stopping heartburn before it starts
Painful results of swallowing air
The mysterious disorder called IBS

One of the most memorable patients in the annals of internal medicine was a small, wiry Irish-American who is known to history simply as Tom. In 1895, when he was nine years old, Tom rashly downed a huge gulp of scalding-hot clam chowder that damaged his esophagus beyond repair. All the doctors could do was to create an opening from the outside into his stomach, through which Tom would have to feed himself. He learned to eat by chewing a mouthful and spitting it into a funnel inserted into the opening, but his infirmity embarrassed him and he never told anyone about it except family and close friends. Proud and feisty, he worked hard at a number of jobs, married and raised a daughter, his life outwardly normal; always he was able to eat in private.

Then in the 1930s, while laboring with a sewer crew, Tom injured the ostomy—the surgically created opening—and for the first time in more than 30 years was forced to seek medical help. Eventually he came to the attention of a young resident at New York Hospital, Dr. Stewart Wolf. Both Dr. Wolf and another physician, Dr. Harold Wolff, were intrigued by questions of the body's response to stress, and because Tom's stomach lining, the gastric mucosa, was clearly visible through the opening, he seemed an ideal subject for observation.

To keep Tom nearby, Dr. Wolf offered him a job as a laboratory assistant. After considerable hesitation Tom accepted the proposal, and for the next 20 years he allowed himself to be examined intensely. The result was a windfall of physiological information that added extraordinary detail

to the preliminary observations made a century earlier by Dr. William Beaumont *(Chapter 1)*.

The most significant of these details, Dr. Wolf later said, helped explain the critical relationship "between emotions and visceral function." In color, wetness and turgidity, Tom's mucosa altered its appearance not just by the day but from minute to minute as his emotions shifted. Pale, dry, thin and sticky when he was frightened or depressed, it turned scarlet and swelled up when he became excited or angry. In so doing it mirrored his facial color: When anger made Tom's complexion turn ruddy, his mucosa brightened and gastric juices flowed copiously. Worry brought the opposite response. One day a staff doctor entered the laboratory to look for some documents that Tom had mislaid. Fearful that he would be fired for negligence, Tom remained silent and motionless, and his mucosa blanched. In a few minutes the doctor located the papers and left; as Tom relaxed, his stomach, said Dr. Wolf, "gradually resumed its former color."

What was taking place in Tom under the fascinated scrutiny of the doctors happens to everyone in the privacy of the body's unseen organs. Why it happens is slowly being figured out. Emotional stress can alter the balance of the hormones—the chemical regulators of body operation—which control, among other things, blood pressure, temperature, breathing rate and, above all, digestion.

Such hormonal changes are initiated by the brain as an automatic response to any kind of stress, pleasant or unpleasant—a forthcoming marriage may be almost as upsetting as a

In this rogue's gallery of gas-producing foods, the offenders are baked beans, onions, cucumbers and cabbage. Gassiest of them all are beans, which are loaded with oligosaccharides— carbohydrates that ferment in the colon, creating gas.

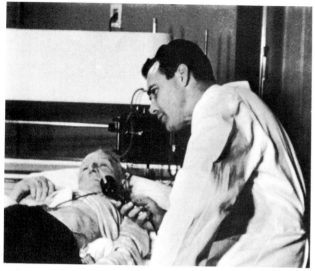

Dr. Stewart Wolf records an experiment in the 1950s with Tom, a patient whose stomach—like that of Alexis St. Martin (page 14) —had been opened by an accident. A balloon connected to a water-filled tube was inserted in Tom's stomach. When Tom grew angry, his stomach contracted, squirting water upward and causing a pen to graph the pressure on a chart.

divorce. Stress is a threat, and the brain prepares a defense by ordering the body to take several steps that will conserve energy for use in combating the threat or escaping from it. Lowering body temperature saves some energy that can be used to respond to the challenge. Increasing blood pressure and breathing rate rushes extra fuel to the muscles. Similarly, digestive activity is slowed or halted—sometimes food is left in place as a partially digested lump, sometimes food is rushed out of the way with diarrhea or nausea and vomiting, sometimes food is simply prevented from entering the system because appetite is shut off.

From pressure and anxiety: queasiness

In all these ways, the anxieties and pressures of daily life can cause gastrointestinal contractions and secretions to undergo rapid shifts in activity. The stomach upsets of students at examination time and of other people in crisis are legendary. The noted internist, Dr. Walter Alvarez of the University of Minnesota, recalled once examining a patient whose stomach was retaining every morsel of a meal consumed six hours earlier: The man had been involved in a bitter political dispute; when he had managed to calm down "his stomach emptied perfectly."

Mild upsets of this kind will often disappear if the patient will take it easy for a day. In most cases the disorder clears up anyway once a special crisis is past. But for a significant minority, the trouble does not go away. The simple stomach-ache becomes chronic. The diarrhea lingers on indefinitely—and infuriatingly.

What causes these annoying conditions to persist, in a large proportion of cases, is continued stress: The emotions remain in turmoil, and the gut continues to churn. Often the stress is so inherent and deep-seated that the sufferer is scarcely aware of its constant presence. But the digestive turmoil it causes can generally be alleviated and often banished entirely if the sufferer will learn how to relax and adjust to stressful situations or will make a few basic compromises in daily habits. Beyond that, there are various medical techniques for relieving the physical symptoms.

Just why some people react to the wear and tear of life more noticeably than others—and why some seem blissfully impervious—is not known. Some individuals have basic physical weaknesses that make them unusually vulnerable. And some may be impervious, researchers at the University of Utah suggested, because they always manage to focus on the nonstressful aspects of any situation.

One group of gastroenterologists reported that of the five most common disorders they treat—the diverse ills called irritable bowel syndrome, ulcers, gall bladder problems, inflammation of the esophagus, and heartburn—only the gall bladder ailments do not seem related to stress. Also blamed on stress are other common afflictions: chronic gas and hemorrhoids. All the disorders except ulcers *(Chapter 4)* cause the sufferer more discomfort than actual physical danger. Yet the discomfort can be so persistent and annoying that it darkens daily life.

Why stomach acid goes astray

One disorder that in many cases is linked to a physical condition is heartburn. For most people it is merely an occasional bother, but for some it can be painful and incessant. It is brought on when stomach acid backs up into the esophagus, causing a burning sensation—either mild or sharp—in the chest. It may last just a few minutes or go on for an hour. Some victims are convinced they have suffered a heart attack. Physicians refer to this backing up as acid reflux, or simply reflux.

Any sudden physical straining or bout of nervousness can bring about heartburn. A simple belch can produce heartburn, the vented air bringing up acid to sear the unprotected esophageal lining. Bending over or lying down will aggravate the condition. A typical sufferer is the executive who returns from a heavy lunch, sits down, leans back and puts his feet up on his desk. He may soon clutch his chest and wonder why the food seems to be traveling the wrong way. Others may be suddenly stricken in the middle of the night.

In many cases, the victim contributes to the ailment by eating too much or too rapidly — often in reaction to nervous stress — and thus so overloads the stomach that a backup is inevitable. People who dine while tense or angry are particularly vulnerable, for these conditions are likely to increase the flow of stomach acid — as Tom many years ago demonstrated. One physician, for example, unable to find an organic cause of a patient's severe heartburn, finally coaxed the man into admitting an extramarital fling — the source of the tension that was upsetting his stomach.

Although tension often brings on heartburn, the condition comes on more readily in people suffering from hiatal hernia, a common abnormality at the juncture between the esophagus and the stomach. At this point, where the esophagus passes through the diaphragm, the opening — or hiatus — may become stretched, allowing the upper end of the stomach to move up through it into the chest cavity. That this shift in position can cause heartburn is owing to the presence there of the lower esophageal sphincter, which is supposed to act as a one-way valve allowing food to travel down but not up. When the sphincter rides upward it no longer is reinforced by the confining tissue of the diaphragm and, thus unsupported, it may be unable to keep gastric acid in the stomach.

Doctors believe that the tendency for a hiatus to enlarge is congenital. What brings on the herniation, though, is the gradual weakening of the sphincter muscle with age; this loss of strength is often aggravated by tight clothes, an accidental blow to the stomach or a strain. Several young men have developed hiatal hernias when they tried to prevent their jacked-up cars from falling.

The trouble with blaming hiatal hernias for all heartburn is that such digestive discomfort is experienced by many persons without hiatal hernias, while many others are herniated without ever knowing it and without being particularly bothered by heartburn. At least half the population over 50 has a hiatal hernia of some kind, and the susceptibility rises as the years progress. Women are more likely to be affected than men, perhaps because of the strains of pregnancy.

Stopping heartburn before it starts

If severe heartburn is an uncommon experience for you, there is no need to alter your habits. But if it occurs with alarming frequency, you should take advantage of the remedies avail-

As his puzzled bride looks on, an unsure bridegroom doubles over with an untimely attack of abdominal agony in this whimsical 19th Century French lithograph entitled "Colic on His Wedding Night." Colic, a broad term for any sharp abdominal cramp, can be caused by physical ailments or, more often, by psychological stress like that afflicting this unfortunate man.

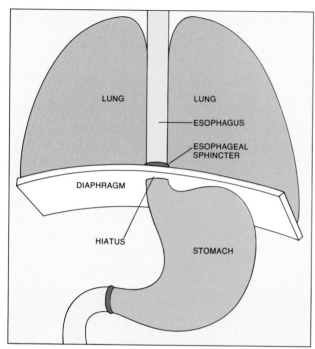
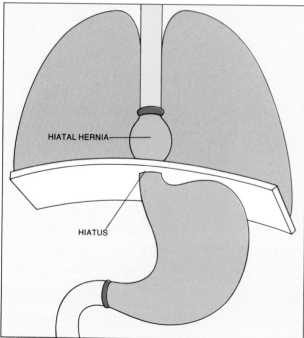

Heartburn often is due to hiatal hernia, sketched above. Normally (top) the stomach remains below the diaphragm, and the esophagus reaches the stomach through an opening, the hiatus. If this opening stretches (bottom), the stomach bulges upward and increases pressure on the sphincter valve, squeezing acidic stomach contents into the sensitive esophagus.

able to allay or even prevent it. Some are mechanical: Wear easy-fitting clothes — constricting the abdomen can force acidic stomach contents upward. Because a contributing factor is simple gravity, avoid lying down within two or three hours after a meal; a resolute verticality may ensure that your stomach empties itself in the proper direction. Raising the head of your bed six or eight inches, which will slant your esophagus enough at night to discourage reflux, also helps — and so will giving up midnight snacks.

A few changes in patterns of consumption may be beneficial. Cut down on smoking and on spicy or fatty foods, chocolate and coffee, all of which have the unexpected effect of reducing the muscle tone of the lower esophageal sphincter, thus encouraging reflux. Carbonated beverages, which stimulate belching, should be shunned, as should foods that contain large amounts of air — whipped cream, soufflés and even milk shakes. Reduce your intake of citrus juices, which are acidic enough to irritate the esophagus directly as they go down. Alcoholic drinks stimulate acid secretion in the stomach, aggravating reflux. Many small meals are better than a few large ones. And whatever food you select, consume it in as pleasant and relaxing a setting as possible.

The more you relax, in fact, the less likely you are to be hit by heartburn. Stress is woven into the texture of contemporary life, from the pressures of the workplace to the vague anxieties that arise from shifting social values, economic uncertainty and political turmoil. The very habits of heavy smoking, compulsive coffee drinking and the consumption of meals on the run are classic reactions to stress. Yet most people find it possible to slow down and take life easier, often simply by telling themselves to do so. In addition, regular hours, moderate habits and regular physical exercise all relieve stress and promote general well-being.

Such preventive steps are unlikely to give total protection. When heartburn arises, you can mitigate it by taking an antacid. One remedy that has proved particularly effective is a combination of alginic acid, which coats the lining of the esophagus on its way down, with three commonly used antacids: aluminum hydroxide, magnesium trisilicate and sodium bicarbonate.

If these ministrations fail, it is time to consult a doctor, who will probably take tests to see if the ailment is a symptom of some larger ill, such as an ulcer. Barring this, the physician may prescribe medicine more potent than drugstore antacids. One effective prescription drug is cimetidine, developed for the treatment of duodenal ulcers but also used by many doctors to ease heartburn.

In a few unlucky souls, acid reflux is so persistent that the lining of the esophagus becomes inflamed, a condition known as esophagitis. The inflamed mucosa may start to bleed, and the swallowing of food becomes a painful enterprise. Often a hiatal hernia is the underlying cause. Under such circumstances, and if examination with an endoscope reveals the lining of the esophagus to be severely eroded, the tear in the diaphragm may have to be repaired surgically.

Painful results of swallowing air

More widespread as a digestive complaint than heartburn but similarly related to stress is an overabundance of gas in the system. Everyone carries a certain amount of gas in the gastrointestinal tract — about two thirds of a cupful on the average at any one time. Of this, only a small portion will normally be found in the stomach, and even less in the small intestine; most resides in the large intestine, the colon. Except for a small amount of oxygen absorbed into the body, none of this gas serves a useful purpose, a fact that makes any momentary increase irksome. As the body keeps acquiring gas it is also constantly seeking ways to vent it, and normally 10 times the basic two-thirds cupful passes out of the system every 24 hours.

Gas may be taken in from outside the body or manufactured within it. A surprising amount arrives from the mouth with every swallow. Saliva itself contains tiny air bubbles, and anything that increases salivation — smoking, chewing gum, sucking on candy — brings in extra air. Carbonated beverages and beer not only carry gulps of air into the stomach, but once there, release bubbles of carbon dioxide. Many foods, such as popovers, whipped cream and meringues, contain more air than solids. Some edibles are unexpectedly airy: A quarter of the volume of an apple is actually air.

Although the body absorbs some of it, much of the swallowed air comes right back out again as a belch. Some cultures consider an exuberant burp after a good meal to be a compliment to the host; the 15th Century Emperor of China, Yung-lo, by one account was capable of 20 belches a minute, a talent that could bring great pleasure to those who served him, ''for the Emperor eructated at that rate only when the gourmet in him was perfectly gratified.''

In some people, however, the swallowing of air—the doctor's word for it is aerophagia—becomes excessive, and the venting problem becomes formidable. The victim feels bloated and cramped. Air that will not come up collects in the upper part of the stomach in a painful giant bubble known as a *magenblase*. That which does come up can bring heartburn. Belching becomes an almost continual problem.

Doctors are convinced that virtually all such excess can be traced to anxiety and nervous tension. One authoritative text blames ''a tic or a habit spasm,'' and goes on to suggest, ''If a physician will ask the patient to belch while he observes him in profile, he will see that the belch is preceded by a forward movement of the head and chin on the neck, that this is associated with a voluntary intake of air, and that the swallowed air is soon expelled.'' Among the simple remedies proposed are ''drinking slowly, avoiding the use of straws, eating calmly, and trying to avoid air swallowing,'' as well as this tip: ''Biting on a cork is said to prevent air swallowing, but many patients learn to swallow even around this obstacle.''

Gas in the lower tract

Even under normal circumstances, of course, not all the air swallowed is directly absorbed into the body or burped up. A small amount is pushed on down the tract, out of the stomach and into the intestines. There it joins other gases produced by the body itself. Some of these gases result from what is known as diffusion, gas molecules seeping into the intestinal tract from the blood in response to changes in pressure within the body; carbon dioxide in particular may be introduced this way—although it can also travel in the other direction, being diffused from the intestines into the blood to be expelled through the lungs. Most intestinal gas, however, is manufactured within the tract. When the digestive juices secreted by the pancreas combine with the stomach's hydrochloric acid for the purpose of breaking down proteins and other substances, one by-product is carbon dioxide. But a greater volume of gas is contributed by hydrogen, produced in the large intestine by bacteria consuming the undigested remains of food. In certain people bacterial action also yields methane—the same inflammable substance that fuels gas cookers and heaters.

Doctors have no idea why some individuals generate methane while others do not. Dr. Michael Levitt, of the Veterans Administration Medical Center in Minneapolis, once studied the intestinal workings of a pair of identical twins; only one twin generated methane.

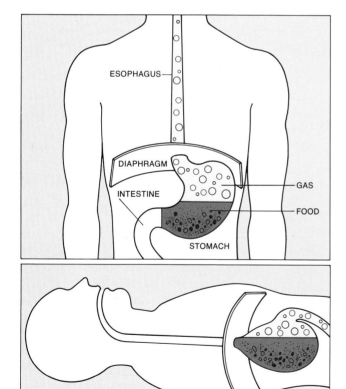

The drawings above show why doctors advise against lying down right after a meal. When the stomach is vertical (top), gas rises to the top and out through the diaphragm and esophagus. When the stomach is horizontal (bottom), liquefied foods may seal off the esophagus, forcing gas into the intestines, where it can cause pain that often is needlessly frightening (opposite).

All of these substances, which account for 99 per cent of gas in the tract as a whole, are odorless. But also generated by bacterial action in the large intestine are small quantities of ammonia and hydrogen sulfide, which is the odoriferous agent in rotten eggs.

Gas inside the tract is also noisy. Everyone's intestinal tract rumbles to a certain degree—a phenomenon that Hippocrates, echoing the sound itself, named *borborygmus*. Caused by peristaltic contractions that push gas and liquid through the canal, the rumbling can change according to the nature of the intestinal activity. A Philadelphia surgeon, Dr. LeRoy Stahlgren, once recorded gastrointestinal sounds as heard through a stethoscope. Under normal circumstances soft, gurgling sounds were heard. If diarrhea was present the sounds were more frequent, though no louder. But if there was intestinal obstruction, the contractions yielded loud, high-pitched noises; and if excess gas was the complaint, there was a series of rushing, crackling noises. Sometimes doctors will diagnose intestinal ailments partly by listening through what Dr. Stahlgren called the "auditory window."

Undeniably, retention of gas in the colon can be bothersome, and often painful. Because the large intestine is shaped like an upside-down U—ascending on the right side of the body, passing across and then turning downward on the left—and because gas tends to rise, the pain is likely to be felt in either of two places. One is the top of the ascending colon, a turning called the hepatic flexure, as it is near the liver. The other is the point where the colon makes its downward turn—called the splenic flexure, as the spleen is nearby. At both points the pain can be misleading. Gas trapped at the hepatic flexure (opposite) is sometimes mistaken for gallstones, while gas at the splenic flexure, below the heart, can mimic the heart ailment angina pectoris.

The foods to avoid

One remedy is to avoid foods that tend to engender gas. The reputation of Boston baked beans for causing flatulence is based on scientific reality, for they contain a complex sugar called stachyose, which passes through the small intestine undigested and, when attacked by the colon's bacteria, releases hydrogen and carbon dioxide. In one test, subjects observing an ordinary diet were switched to one in which they got 57 per cent of their calories from pork and beans; their flatus multiplied 12 times. Other types of beans also foment gas, but to a lesser extent. A high-fiber diet causes an increase in gas; the cellulose arrives in the large intestine undigested and, like stachyose, is attacked by bacteria.

One class of foodstuffs that may cause gas in some groups of people is made up of milk and milk products. The sugar in milk, lactose, must be broken down in the small intestine by the enzyme lactase into two other kinds of sugar, glucose and galactose, before the nutrient components can be absorbed. But tremendous numbers of adults are deficient in lactase. Most blacks, Orientals and American Indians are, in the medical term, lactose-intolerant; so are more than 60 per cent of all Israeli Jews. Among people of Northern European ancestry the frequency is lower—about 10 to 15 per cent. To varying degrees, such individuals suffer from excess gas and, in more severe cases, from diarrhea and cramping if they consume milk products.

If you suspect that a lactose intolerance may have rendered you susceptible to excess gas, there is a straightforward way to make a crude test. Simply give up all milk products—including processed foods, such as breads, made with "nonfat milk solids"—for several weeks. If your symptoms disappear, and stay away so long as you avoid milk products, you may be lactose-intolerant. This kind of self-testing can hardly provide definitive proof, however, because psychological effects have such an immense influence on digestive distress; it is very easy to become convinced that a particular foodstuff—milk, shellfish, eggs or whatever—causes illness when that item actually has no impact on the body, only on the mind. For this reason, many authorities believe that the incidence of true lactose intolerance may be substantially overstated.

If lactose intolerance is physiological rather than psychological, milk products may not have to be renounced altogether. Many foodstuffs made from milk end up with little or no lactose in them. The lactose in butter is broken down in the manufacturing process, and buttermilk and yogurt contain little lactose. Cheeses vary: Among those that are low in lactose are cottage cheese, ricotta and hard cheeses such as cheddar. As for milk itself, a lactase additive is available at pharmacies, health food stores and even supermarkets; poured into a container of milk, it breaks down the lactose within 24 to 48 hours so that the milk can be drunk without ill effects. If none of these foods can be tolerated and you must forswear milk altogether, your doctor will probably prescribe calcium pills—this mineral is essential, and milk is its principal natural source.

Diet changes and measures to ease stress generally bring relief from chronic gas pains. No drug is a proved antidote: Simethicone, a frequent additive to commercial antacids, is reputed to cut down flatus by causing smaller air bubbles to coalesce and form larger air masses that may be more readi-

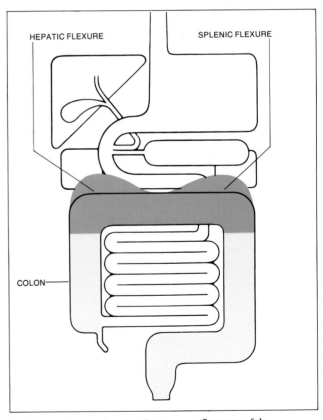

Gas is sometimes trapped at the turns, or flexures, of the large intestine (above), swelling the colon and causing pain: The condition is harmless but mimics the symptoms of serious diseases. Distension of the hepatic flexure can be misdiagnosed as gall bladder trouble, while swelling of the splenic flexure can cause radiating chest pains often mistaken for a heart attack.

ly vented, but doctors are undecided about its effectiveness.

A number of mechanical maneuvers can be tried to move gas that is trapped in the colon. Pressure can be applied to the abdomen either by kneading it manually or by lying on the stomach with a pillow under the hips. Sitting on the floor with knees drawn up and rocking back and forth may help. In the event of severe cramping, some doctors suggest lying down on the side that hurts, drawing up the opposite knee, and rocking back and forth. Finally, Dr. Levitt remarked, ''A brisk walk or a jog appears beneficial in some persons,'' because heightened muscle tone encourages the body to relieve the pressure.

The mysterious disorder called IBS

Although too much gas is usually a disorder unto itself, it can also be a component of one of the most perplexing of all the ailments associated with the digestive tract. This is the chronic and mysterious disorder that has been called spastic colon, colon neurosis, nervous diarrhea, nervous stomach and mucous colitis, but now goes by the catch-all title of irritable bowel syndrome, or IBS. It has no distinguishing organic cause. All tests tend to show up negative. There is no inflammation of the gut, no obstruction, no diseased tissue.

Even so, IBS can be agonizing, and it plagues an extraordinary number of people. Reputed to be the most frequent cause of industrial absenteeism after the common cold, it is the commonest gastrointestinal disorder that brings people to see a doctor. One guess is that 22 million Americans suffer from it. Those who fall prey to IBS tend to be young when first afflicted: Onset generally occurs in the 20s, or even in the teens. It afflicts more than twice as many women as men. The sponsors of a call-in service, called Gutline and operated by the American Digestive Disease Society, report that IBS is the main subject of calls.

Victims of IBS have diarrhea, or constipation, or the two complaints may alternate. They often feel gassy. Some have pain, others not. Said one, ''I have a pain in my chest sometimes so sharp I can't move. Sometimes I get heartburn. Other times I get very bloated. I'm constipated too. Medicine doesn't seem to help.''

IBS seems to be related to an affliction common in past centuries, a feminine disorder called the vapors, about which the irrepressible Benjamin Franklin had much to say. ''There's no Disease puzzles Physicians more,'' he wrote, than this complex of symptoms that included ''a Croaking of the Guts, a Fulness of the Stomach,'' and something that sounded very much like heartburn. At the same time the sufferer would experience ''a great Heaviness and Dejection of Spirit, and a Cloud seems to hang upon all her Senses. In one Word, she has no Relish for any thing, but is continually out of Humor, she knows not why, and out of Order, she knows not where.''

Today's IBS patients often freely acknowledge a psychological component to their ailment. ''I am an individual who gets involved in too many projects,'' reported one victim, ''and am always on the go. I know this is very bad for my condition. There are times I am free of all symptoms, and feel fine. However, my symptoms are always just around the corner, waiting to catch up with me.''

Another factor may be an inability to express strong emotions. The illness occurs, said Dr. Henry Maudsley, an English psychiatrist, when ''sorrow has no vent in tears but may make other organs weak.'' Perhaps in IBS the colon figuratively weeps. Those who agree with this argument point to statistics that show IBS sufferers to be more neurotic and anxious than most and suggest IBS may be largely a psychosomatic ailment: In one survey, 78 per cent of patients with IBS were found to have experienced some form of psychiatric illness, while only 15 to 25 per cent of non-IBS sufferers had such a record.

Although IBS may be psychosomatic, it is not imaginary. The pain and discomfort are real. And there is cause for concern: Some IBS symptoms resemble those of cancer, as well as of other serious diseases, including liver trouble and inflammation of the intestines (Chapter 5). Because IBS does not show up on standard clinical tests, there is no easy diagnosis. But if laboratory tests, interviews and physical examinations rule out every other ailment, a significant event occurs in an overwhelming number of cases. Because the patient has been so worried about organic, crippling disease, the removal of that concern immediately reduces stress, and there is likely to be a marked easing of IBS. "When I tell the patient he doesn't have cancer," said one gastroenterologist, "he begins to get better right away."

At the same time, the doctor will be careful to make plain that IBS is a condition for which medicine has no real cure. Not that medicine has neglected to try. Ben Franklin, echoing the standard therapy of his day, recommended a purge, an emetic and a spartan regimen to banish the complaint:

Endeavour to preserve a cheerful Spirit, putting the best Construction upon every Body's Words, and Behavior: Plunge, 3 Mornings every Week, into cold Water, over Head and Ears; or if you have no Convenience for that, it will have the same Effect, if you suffer your self to be whip'd with smart little Rods. It can't be imagin'd how this will brace the Nerves.

The contemporary view tends more toward quiet accommodation to the inevitable. "This is an incurable, chronic, relapsing disease," declared Dr. Albert Mendeloff of Johns Hopkins, who went on to suggest that the best therapy is learning to deal with the stresses of daily life. Dr. Walter Alvarez agreed: "A woman must be made to see that she must accept her colon pretty much as it is, and she must face the fact that she is likely to have some distress in it off and on for the rest of her days."

That is no reason for gloom, however. With careful management, the symptoms can be made to recede markedly. Most gastroenterologists start by recommending some dietary changes, in particular an increase in roughage. Fiber-rich foods such as bran, whole grains, green vegetables and fruit are effective antidotes to constipation.

Over the weeks the doctor may suggest further changes— perhaps cutting down on sugar, alcohol, soft drinks and beef, all of which may exacerbate IBS symptoms. He may prescribe a mild tranquilizer or antidepressant; some physicians try an anticholinergic, a medicine that impedes the muscle spasms of diarrhea.

Sometimes the patient can steer clear of the source of stress, though this is patently difficult in the modern world. Remarked Dr. Marvin Schuster, of Johns Hopkins, "There are certain stresses which a person at particular points in his life cannot tolerate and ought to avoid. A patient may for instance not be able to accept a job promotion that puts him into a situation that is extremely stressful. He has to arrive at ways of handling these things." Some physicians have found success with muscle-relaxation techniques, such as biofeedback, in which the patient learns to control muscle tension. Meditation techniques may also help.

Causes of hemorrhoids—and the cures

One frequent consequence of irritable bowel syndrome, because of the cumulative effects of both diarrhea and constipation, is hemorrhoids, or piles. They too tend to be chronic. And while the cause is sometimes physical, they can also be stress-related insofar as they result from repeated—and generally unnecessary—straining at stool.

Hemorrhoids consist of clusters of dilated veins in the lower rectum and anus, and they are a concern to untold numbers of people. More than half of all men and women over the age of 30 have hemorrhoids. Each year these sufferers buy more than 55 million dollars' worth of nonprescription medicines to alleviate their distress.

Many victims fear that the disorder is a sign of oncoming cancer. Such fears are generally groundless; cancer rarely occurs around the anus. On the contrary, hemorrhoids, although often painful, are usually a minor affliction that can be relieved easily and quite often cured completely.

Although heredity may influence their occurrence—some families seem more disposed to contract them than others— they tend to be caused by specific habits and activities. The straining that is a hallmark of constipation is perhaps the commonest cause. Some people's occupations may contribute to the condition; those who must constantly stand (department store salesclerks) or lift heavy objects (stevedores) can put abnormal pressure on the lower pelvic area, while those who continually sit (truck and bus drivers) may paradoxically produce the same effect. Pregnant women often develop hemorrhoids because of pressure on the rectal and anal veins, but the condition generally improves after the birth of the baby. Sometimes hemorrhoids occur as a side effect of more serious diseases, such as cirrhosis of the liver or heart failure.

Although the basic cause of hemorrhoids may be difficult to identify, the course of development is plain. A group of veins, dilated by pressure so that they are continually stretched, eventually cease to return to normal. They thereupon remain, as swollen tissue just inside the anus (internal hemorrhoids) or just outside (external hemorrhoids). Straining will often cause internal ones to protrude to the exterior, where they cause discomfort; external hemorrhoids may be almost constantly bothersome and often painful. Internal hemorrhoids in particular may bleed, especially during a bowel movement. In addition to hurting, they can itch.

Most cases of hemorrhoids respond to treatment by the patient— though persistent attacks or the presence of blood in the stool may call for a doctor's attention. Avoiding any kind of physical straining will alleviate the pressures that generally cause the condition. Using one of the stool softeners described in Chapter 2 may help, as will the habits advised for the treatment of constipation. Hot sitz baths— three or four times a day, for 20 minutes at a time if possible— can be immensely soothing. The affected area should be kept clean with soap and warm water.

In most cases such procedures will banish the discomfort in a week or two and the veins, while remaining more or less swollen, will cease to annoy. Often internal hemorrhoids that have protruded can be pushed back manually, easing the symptoms greatly.

Ointments can lessen pain, inflammation and itching. Ordinary petroleum jelly works well, as do the many hemorrhoid preparations available in drugstores. Although none cures hemorrhoids, or shrinks hemorrhoidal tissue except temporarily, they can offer temporary respite while the swelling disappears spontaneously.

The basic ingredient of all preparations is a lubricant or protectant, to provide a coating over inflamed tissues and reduce itching. The substances used include lanolin, zinc oxide, cocoa butter, cod-liver oil and many others; one product contains shark-liver oil. All work acceptably, according to a government review panel. Another common ingredient is a local anesthetic such as benzocaine or pramoxine hydro-

chloride. These can have a mild pain-relieving effect on external hemorrhoids, although continued use of either can eventually oversensitize the affected tissue. Also, some people will develop an allergy to benzocaine.

Some preparations contain vasoconstrictors, which shrink tiny blood vessels. Among them are ephedrine sulfate, epinephrine hydrochloride and phenylephrine hydrochloride. The chief argument against them is that hemorrhoidal bleeding comes from large blood vessels rather than small, and so the vasoconstrictors may be superfluous. The official panel investigating hemorrhoidal preparations approved their use but advised that they should be avoided by anyone who is taking tranquilizers or who has heart disease, high blood pressure, hyperthyroidism or diabetes. There is no proof that astringents, such as the witch hazel or tannic acid included in some remedies, constrict tissue, but they are harmless.

Many preparations are available as suppositories instead of, or in addition to, ointments. Their effectiveness is disputed by some doctors, who point out that the suppository is likely to pass right by the inflamed area and melt where it is not needed. They also note that pain and itching are less likely to stem from internal hemorrhoids than from external ones, which the suppositories do not affect. Nevertheless many patients claim to be helped by them.

There is one overriding remedy available to persons whose hemorrhoids persist or become difficult to manage successfully: surgery. Permanent removal of the distended tissue is the one real cure *(Chapter 6)*.

The majority of people who have the condition, however, will not need to face that decision. Hemorrhoids, like all chronic, stress-related ailments, are best treated by minimizing the pressures and anxieties that bring them on. How that is done varies with the individual. Said Dr. Nicholas Hightower of the Scott and White Clinic in Temple, Texas, "I tell my patients that the digestive system is like skin. Some of us are born with very sensitive complexions and have to stay out of the sun and watch what soap we use. Similarly, some of us inherit sensitive digestive systems which tend to react strongly to stimuli. It's simply a case of learning to live with one of the body's idiosyncrasies." ✹

Menus to soothe a troubled digestion

A 19th Century medical journal summed up the current advice on digestion in a simple admonition: "Avoid anything which obviously deranges the stomach." The maxim still holds, but obeying it is not easy. Identifying the food that causes a problem is tricky, and even if the offender can be isolated, eliminating it from the diet may bring a new problem by depriving the body of an essential nutrient. The five sample menus at right and on the following pages, prepared by Dr. Howard Spiro of Yale, are planned to provide healthful nutrition while omitting various foods that are known or believed to be responsible for common digestive problems. The menus are samples only; similar foods can be substituted to provide variety or to suit individual tastes.

Two of these diets are based on a clearly established link between an ailment and a chemical ingredient in a food. One ailment involves milk. Many people of African, Oriental or Mediterranean ancestry lack a chemical activator, or enzyme, that digests the sugar in cows' milk, lactose. They get sick to their stomachs when they eat dairy products and can eat very little of the five worst offenders—milk, ice cream, and American, ricotta and cottage cheeses. Some victims must exclude all milky foods and the many processed foods, including most breads, that are made with what the label specifies as "nonfat milk solids."

Similarly, other people must avoid common breads and grain products containing gluten, the sticky protein that enables wheat and rye dough to rise during baking; gluten somehow attacks such a person's intestinal villi, the tiny projections that absorb and process nutrients. Gluten also turns up unexpectedly in processed foods—even in hot dogs, for instance—and victims of gluten intolerance must check package labels.

More controversial are the high-fiber, low-fiber and bland diets; there is little scientific proof that any of these three regimens brings direct physiological benefits. Nevertheless, Dr. Spiro found that the low-fiber diet alleviates diarrhea, that the high-fiber regimen helps to relieve constipation, and that the bland diet soothes patients complaining of heartburn and indigestion. He did not rule out the possibility that the diets work psychologically, helping because the patient believes in them. And to doubts about the bland diet's emphasis on milk—which is known to stimulate the production of irritating stomach acid—Dr. Spiro responded, "Can anything which makes you feel so much better really be so bad for you?"

BLAND DIET

BREAKFAST
Apple juice
Farina
Milk

MIDMORNING SNACK
Milk
White toast
Soft-boiled egg

LUNCH
Cream of celery soup
Tuna fish
Green beans
Milk
Orange juice

MIDAFTERNOON SNACK
White bread
Banana
Egg custard

DINNER
Cream of mushroom soup
Broiled chicken
Noodles
Cooked spinach
Milk

BEDTIME SNACK
Milk
White bread
Flavored gelatin

For sufferers from simple indigestion, this menu excludes all foods thought to irritate the gastric system; coarse-grained breads, fruits and nuts with skins or seeds, caffeine-containing beverages, alcohol and all condiments, spices and seasonings except salt. Frequent small meals, six a day, help control gastric acidity by keeping the stomach full of food.

LACTOSE-FREE DIET

BREAKFAST
Orange juice
Dry cereal
Fried egg
Hard roll or French bread with margarine
Coffee with nondairy creamer

LUNCH
Roast turkey on hard roll with
lettuce, tomato and mayonnaise
Spinach salad
Peach
Coffee with nondairy creamer

DINNER
Tomato juice
Roast beef
Baked potato
Broccoli
Apricots
French bread with margarine
Coffee with nondairy creamer

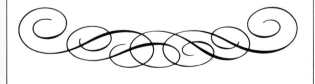

GLUTEN-FREE DIET

BREAKFAST
Half grapefruit
Corn flakes
Scrambled egg
Rice bread with butter
Coffee, tea or milk

LUNCH
Cheese omelet
French fried potatoes
Green beans
Green salad
Apple
Corn bread with butter
Coffee, tea or milk

DINNER
Roast veal
Rice
Cauliflower
Tomato salad
Tapioca pudding
Soybean wafers with butter
Coffee, tea or milk

To make up for the loss of certain nutrients in milk, a diet excluding milk sugar, or lactose, lists cereals fortified with B vitamins; breads, such as traditional French loaves, that contain no milk; and fruits rich in vitamin A. However, dairy products are the only good source of the essential mineral calcium; anyone on a lactose-free diet must take calcium pills.

This menu excludes all grains—wheat, rye, barley and oats— that contain the sometimes troublesome proteins called gluten. Eliminated from the diet are most breads, cereals, pastries and beer. Because gluten is not found in potatoes, soybeans, rice, corn or tapioca, the breads, breakfast cereals and side dishes made from them can provide the carbohydrates needed for health.

HIGH-FIBER DIET

BREAKFAST
Prunes
Bran cereal
Soft-boiled egg
Whole-wheat toast with butter
Coffee, tea or milk

LUNCH
Ham and cheese sandwich on whole-wheat bread
with one small tomato
Broccoli
Blueberry pie
Coffee, tea or milk

DINNER
Roast beef and gravy
Mashed potato
Cauliflower
Watermelon
Whole-wheat roll with butter
Coffee, tea or milk

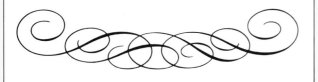

LOW-FIBER DIET

BREAKFAST
Orange juice
Soft-boiled egg
White toast with butter
Corn flakes
Coffee, tea or milk

LUNCH
Chicken noodle soup
Hamburger on bun
Cooked carrots
Banana
Coffee, tea or milk

DINNER
Tender broiled steak
Baked potato without skin
Cooked spinach
Canned pear halves
Rye bread with butter
Coffee, tea or milk

Constipation is helped by fiber-rich whole-grain breads and cereals, berries, dried fruits, watermelon, broccoli and cauliflower. The fiber absorbs many times its weight in water, helping to soften food residues for easier elimination. Sudden increases in fiber intake, however, can cause cramps and gas; fiber-rich foods should be added to the diet over several weeks.

A mirror image of the high-fiber diet, a low-fiber menu can ease diarrhea. In addition to fruits low in fiber, such as bananas, peaches and cantaloupes, it includes some high-fiber fruits and vegetables, which may be eaten provided they are peeled and cooked, as most canned foods are. Tough meats, full of indigestible connective tissue, are to be avoided.

CHAPTER 4

The ulcer enigmas

Myth of the executive disease
The dangerous stress: fast decisions
Does diet make a difference?
Signs that mean life or death
Two revolutionary drugs
New challenges to the milk prescription
How to live with a chronic ulcer

Ulcers, the storied ailment so often blamed on today's pressure-cooker civilization, abound in mysteries. Scientists do not know why more men than women now contract these painful sores in the digestive tract—although the opposite has been true in the past—why they seem to run in families, why the flare-ups occur mainly in the spring and fall, or why an ulcer hurts the way it does. And although there are countless speculations, no one really knows what causes an ulcer in the first place.

Even more bewildering is the fact that ulcer disease—for reasons as yet undetermined—is declining in much of the world. It is not just that individual cases tend to be less disabling than they once were. No mystery there: Some astonishing medical breakthroughs—new tools for early diagnosis, and new drugs that bring seemingly miraculous relief—have saved lives, almost eliminated the need for surgery and enabled most sufferers to control their ailment with a minimum of discomfort. What puzzles doctors is a dramatic fall-off in the total number of cases. The United States Air Force counted 2,005 ulcer cases among its patients in 1967, only 832 cases in 1971; a study in England and Wales estimated 65,189 cases from 1959 to 1962 and only 50,591 from 1971 to 1974.

The decrease in cases is as baffling as the question of what caused most ulcers in the first place. Until the 20th Century, ulcers that attacked the lining of the digestive tract occurred mainly in the stomach. This gastric variety had been observed in ancient Greece. In 19th Century America, it seems to have been confined primarily to upper-class young women, although perhaps the reasons had more to do with the vagaries of historical record keeping than with medical fact. As time went on another kind of ulcer began appearing, in the first section of the small intestine, the part known as the duodenum. Duodenal ulcers were rare at first; only 70 cases were identified in a major English hospital during the entire 19th Century.

Around the turn of the century, however, duodenal ulcers began a mysterious rise throughout the Western world. By World War II they were recognized as a major intestinal disorder. Young females did not often get them—and were not coming down with gastric ulcers much any more, either, to the puzzlement of doctors. Instead, the most frequent victims seemed to be young or middle-aged men. Wartime Britain experienced a virtual epidemic. In 1946 one American authority, Dr. Andrew C. Ivy of Northwestern University Medical School, estimated that one out of every 10 American males would harbor a duodenal ulcer by age 65. Gastric ulcers were up, too, though less so. By the mid-1950s the combined loss of income and medical expense suffered by ulcer victims in the United States alone was almost $500 million a year and rising.

So prevalent were ulcers—and so fixed in the public mind as the unfortunate wages of high-pressure living—that they were seen to be an almost mandatory badge of worldly achievement. It seemed as if everyone who mattered was developing them—President Harry S. Truman, Elizabeth

Tensed over his radar screen as he monitors a maze of flight paths, this air-traffic controller is, according to one study, three times as likely to suffer an ulcer as the population at large. But there is no conclusive evidence tying ulcers to job-related stress alone; other factors such as body chemistry, heredity, life style and personality must usually share the blame.

Taylor, Generalissimo Francisco Franco, and any number of hard-driving, hard-drinking lawyers, stockbrokers and upwardly mobile corporate executives in their thirties and forties. The two status requisites in the advertising agencies of New York's Madison Avenue were a double ulcer and a bottomless capacity for double martinis—or so it seemed.

Like all oversimplifications, this picture was only marginally true. Fully as many executives gloried and prospered healthily under competitive pressure as became ulcerated. And untold numbers of nonexecutives in all parts of the world were coming down with ulcers—factory workers, farmers, stenographers, even teenagers. The roles of stress, living standards, occupations and social milieu in causing this disease—an affliction once called by the great diagnostician Sir William Osler "the wound stripe of Western civilization"—are now much disputed.

Certainly stress and tension have not disappeared from modern society, but in the decades following World War II the tide of ulcers quietly and unaccountably reversed. The number of cases began slowly to ebb. In England by 1968 the annual rate of incidence was only half what it had been a decade before. Five hospitals in Seattle reported in 1977 that the number of ulcer cases they were treating had dropped by 35 per cent in 10 years. More recently, a projected study of ulcer surgery planned at 14 Veterans Administration hospitals had to be called off because not enough patients were available to make up a valid sampling. And while gastroenterologists were wondering what had happened to the vast numbers of ulcer patients that used to crowd their offices, they were seeing a perplexing increase in patients suffering from various mystifying disorders of the lower bowel *(Chapters 3 and 5)*.

"The irony," said New York gastroenterologist Nathaniel Cohen, "is that the decline has come about just when we have really learned how to treat ulcers successfully. The new therapies are showing impressive results." Joked Dr. Morton Grossman of Los Angeles, "The medical profession has to hurry up and solve all the questions about ulcers so that we can take credit for the decline."

There is a long way to go, however. The questions and mysteries linger on—and ulcer disease, despite its diminution, is still a malady of major proportions. If no longer, in the words of Dr. Albert Mendeloff of Johns Hopkins University, "one of the great endemics of the 20th Century," it is still a damaging disorder that continues to vex more than 20 million Americans. Every day, reported the American Digestive Disease Society, another 4,000 Americans develop an ulcer, and the ailment causes a billion-dollar annual drain on the national economy. If ulcer jokes have gone out of style, the illness remains painful, restrictive and potentially very dangerous. If not treated properly and in time, it can be fatal: The silent-movie star Rudolph Valentino died of a bleeding ulcer, and during the early 1980s as many as 6,000 Americans a year succumbed to similar complications. The ailment can never really be cured: Once you have experienced an ulcer you are forever vulnerable, and the recurrence rate is significantly above 50 per cent.

Who gets which ulcer

Part of the ulcer enigma lies in the frequent confusion between the two varieties, gastric and duodenal. The two often afflict different types of individuals, seem to be caused by different physiological imbalances, frequently bring different symptoms and respond differently to treatment. In both manifestations, the ulcer is physically the same: a festering open sore resembling a small volcanic crater in the wall of the digestive tract. (The entire body, in fact, is susceptible to similar kinds of ulceration, inside and out: in a blood vessel, on the skin or even in the eye.)

The ulcers in the digestive tract, both gastric and duodenal, begin as mere pinpoints of erosion and then increase in size, commonly spreading to as much as three quarters of an inch or more across. Some have been measured at a width of three inches. Both types are caused by the corrosive effect of gastric juices on the lining of the stomach or the duodenum—and sometimes, though rarely, on the lining of the esophagus. The juices, in effect, begin to digest the stomach or duodenum itself.

In a normal digestive system, both stomach and duodenum are amply protected against gastric secretions. As food is consumed, the brain signals the stomach via the vagus nerves to begin secreting two activating chemicals, a hormone called gastrin, and the multipurpose body chemical histamine. Together they cause cells in the stomach wall to produce two key ingredients of the digestive juices. One is hydrochloric acid. The other is pepsinogen, which is itself harmless but which when activated by the hydrochloric acid is turned into pepsin, a corrosive compound. Scientists believe that pepsin does the actual digesting; the hydrochloric acid provides the necessary environment and stimulus.

Acid is necessary for some steps in digestion, and the opposite—an alkaline state—is necessary for others. The system must control the acid-alkaline balance, shifting it one way and then another with a bewildering array of hormones, secreting cells and nerves. Most foods are neutral or slightly acid when they enter the tract. When they reach the stomach, the release of hydrochloric acid produces a mixture as acid as undiluted vinegar—3.0 on the scientists' pH scale of acidity, in which 7.0 is neutral and numbers above 7.0 indicate increasing alkalinity. The acidity of the stomach mixture is quickly neutralized in the small intestine by sodium bicarbonate in the pancreatic juices—pH 8.0. In the large intestine the balance shifts again to provide the slightly alkaline environment required by the bacteria that decompose undigested fiber; in the cecum, for example, the pH reaches 7.3.

The strong acid in the stomach is needed to process proteins, abundant in meat; other types of food—carbohydrates and fats—do not require acid for digestion. That the acidic mix of pepsin and hydrochloric acid does not instantly begin eating away at the stomach wall is due to the protective effect of the mucus and other substances that are constantly bathing the wall even as the pepsinogen and hydrochloric acid are being secreted. So the stomach wall not only produces the potent digestive juices, but also maintains the defense mechanism that will protect it against self-damage.

Hazards of a weakness in the stomach wall
Trouble occurs when the tissues of the wall fail in their normal protective role, or when the vagus nerves order so much digestive secretion that the tissues are overwhelmed. It is

A baffling decline

No one knows just why, but ulcers seem to be going away. The decline during the 1970s, continuing a generation-long trend, is documented in figures on the numbers of people hospitalized with ulcers *(below)* and the numbers who died from the disease *(bottom)*. Both deaths and hospitalizations for men were reduced by almost half over the decade in the United States, but American women did not fare nearly so well. One explanation for this male-female difference may be social changes that have heaped ulcer-provoking stresses on women, narrowing the once-sizable gap in incidence between the sexes.

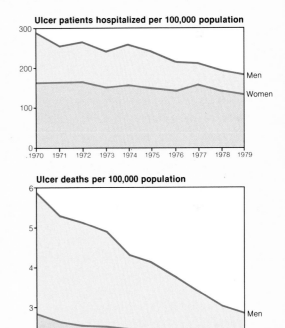

Ulcer patients hospitalized per 100,000 population

Ulcer deaths per 100,000 population

* Estimated for 1979

gastric ulcers that usually result from the first of these events. In some people the stomach wall eventually becomes less resilient. Perhaps it generates less of the protective mucus. Doctors are not sure how this occurs; presumably there are many causes.

One way the stomach wall becomes weakened is through inflammation—the condition known as gastritis. Although it does not necessarily bring about an ulcer, gastritis may damage the lining just enough to make it unduly vulnerable to stomach acid. Alcohol and aspirin sometimes cause gastritis; so too may a misdirected flow of liver bile. Normally, of course, bile does not touch the stomach at all; entering the digestive tract by way of the duodenum, it handily aids in the digestion of fats still present in the food at that point. From there it flows harmlessly the rest of the way down the intestines. But under certain circumstances, the bile can back up through the valve connecting stomach and intestines, the pyloric sphincter, and enter the stomach, where it irritates the wall. If a gastric ulcer ensues, in most cases it will develop just above the pyloric sphincter, on the so-called lesser curvature of the stomach's upper wall.

Although gastric ulcers are presumably caused by acid, people who contract them do not ordinarily generate a measurable excess of the fluid. But because the tissues of their stomach lining have been damaged, a perfectly normal amount of acid becomes destructive. Victims of duodenal ulcer, on the other hand, do tend to be burdened with too much acid. Some simply generate an overabundant flow of it: Their stomach lining possesses too many acid-secreting cells for a proper balance. In other sufferers, the secreting mechanism, which is usually operated by the vagus nerves, fails to shut off when it should.

In a healthy system, the cells in the stomach will sense the level of acidity there, and when that level has reached a certain point they automatically shut down the production of gastrin, the instigator of acid flow. This is fortunate, for otherwise the mere thought of food might cause a deluge of acid; as Yale University gastroenterologist Howard Spiro put it, ''the stomach of the gourmet would always be secreting, and all the friends of Escoffier would have to eat mush'' to absorb the excess. But in some cases, acid production may go beyond the normal cut-off point, because of continued signals from the vagus nerves. Such protracted signals are most often provoked by stress or emotion. In some people, moods of anxiety or sorrow can provoke a continual flow of damaging amounts of acid.

Telling the ulcers apart

So long as the stomach wall is healthy, even a surplus of acid will not damage it. The excess acid passes through the pyloric sphincter and pours into the duodenum. It wreaks its mischief most often on the first inch or two, just beyond the sphincter—the site of virtually all duodenal ulcers. Into this short stretch, just before the duodenum turns sharply downward toward the rest of the small intestine, the hyperacidic chyme of ground-up food squirts repeatedly out of the stomach with devastating effect.

''Imagine a stream of schoolboys,'' wrote Liverpool physician Robert Kemp of this spurting stream of corrosive material, ''rushing out of the classroom through a swing door into a short vestibule with a sharp right-angled bend. Everyone of these schoolboys hits the facing wall in the same spot and, while the classroom is in use, will continue to do so.'' The effect is compounded by the fact that the duodenal lining, unlike the lining of the stomach, is not well equipped to defend itself against acidity.

These distinctions of acid levels and tissue durability mark the differences between the victims of the two types of ulcer. Most of those stricken by gastric ulcers generate normal amounts of acid but possess weakened and vulnerable stomach linings; those with duodenal ulcers generate too much acid and inflict it not on the stomach but on the first few inches of the small intestine. There is one other key difference. A gastric-ulcer sufferer tends to keep food in his stomach too long—in many cases the pyloric sphincter fails to open adequately, or the stomach muscles are too weak to push the food through—so that the acid has a longer period in which to work, and it works on the stomach lining. By contrast, a duodenal victim may empty his stomach too soon, so that the duodenum bears the onslaught of acid that has been only partially neutralized by the reactions of digestion.

Although the two classes of ulcers arise in different ways and in different locations, their impact on the tissues of the digestive tract is the same. The linings of both stomach and duodenum are made up of several layers *(page 23)*. A tiny, incipient ulcer may affect only the first, mucus-covered layer and hardly be noticed. But if allowed to fester, it will eat down into the lining and begin making itself known. If it reaches the layer that contains blood vessels, it may cause bleeding. If the ulcer penetrates all the way through, it will open a hole into the abdominal cavity, which invites infection of the entire abdominal lining—peritonitis—and is bad news indeed, requiring immediate hospitalization.

At each stage in the development of either kind of ulcer, there is likely to be pain. The pain differs from most other varieties of intestinal distress in that it tends to occur like clockwork in response to the processing of food. As food and drink are swallowed, acid and pepsin gush out into the stomach in great quantities—but the newly arrived foodstuffs absorb them and buffer their action, keeping overall acidity within bounds. For many sufferers the agony begins after the processing has been completed and the resulting chyme has passed along to the next stop: Leftover acid strikes the now-unoccupied tissue of the stomach or duodenum. Gastric ulcer victims typically begin to feel pain about half an hour after completing a meal; duodenal sufferers are often spared until two or three hours have passed.

Ulcer pain is not sharp. It is usually felt as a gnawing or aching sensation, like an acute hunger pang, at the site of the ulcer itself. It may last anywhere from 30 minutes to a couple

An ulcer appears as a crater-shaped sore in this magnified cross section. The ulcer has eaten through the top layers of tissue, the lacy mucosa and submucosa, and into uniformly dense muscle underneath. Under the dead tissue in the crater, a darker crescent of blood vessels and repair cells forms a healing scar.

of hours, then disappear completely as the acid secretions abate. New York surgeon M. Michael Eisenberg, who has written extensively about ulcers, described the pattern of pain suffered by a Minneapolis bank employee. His duodenal ulcer gave him pain in the morning just as he settled down to work. A snack caused the pain to moderate: The acid was being kept busy. His ulcer hurt again during the afternoon. But the most vexing time came during the night: Even if he had eaten a precautionary snack before retiring, he usually awakened in agony a couple of hours later. In the morning before breakfast he felt fine.

After a few weeks of such attacks, ulcer pain may unaccountably vanish for several months, then abruptly recur. The attacks seem to take place with amazing predictability in either the spring or the fall, though no one can say why. ''Medical science,'' wrote gastroenterologist Burrill Crohn, ''has devoted much time to the study of barometric pressures, changes of moisture in the air, changes in temperature, concentration of dust in the atmosphere, etc. While volumes have been written on this interesting study, we still do not understand why just about Easter time in the spring, and just about late September or early October in the fall, ulcers have a tendency to reopen and show active symptoms.'' Dr. Nathaniel Cohen of Mount Sinai Hospital in New York City said that his ulcer patients, when told that their seasonal pain pattern is so pervasively predictable, come up with all sorts of imaginative explanations, from shifts in the earth's rotation to changes in available foodstuffs to ''tides of psychosomatic stress.'' But nobody really knows.

Almost as obscure is what causes ulcers to hurt at all. The tissues lining both stomach and duodenum contain no nerves for sensing pain. If you were to drip a bit of acid on a cut in the skin anywhere on your body, it would hurt terribly—but if you could somehow jab the lining of your stomach with a needle, or burn yourself there, you would feel virtually nothing. One guess is that the gastric acid bathing a raw ulcer in the stomach wall may cause local muscles to go into spasm: The pain may be the result of a muscle cramp. But there is no proof—just another ulcer puzzle.

Capricious and cryptic, ulcer pain is very real to anyone who feels it. And it often gives rise to a widespread and abiding concern: Does it signal something worse, perhaps cancer? Indeed, does having an ulcer increase the likelihood of developing cancer? The second question is easier to answer than the first: rarely, and in the case of duodenal ulcer, virtually never. Ulcer and cancer are two distinct and dissimilar phenomena, with different causes and different ways of developing, and an ulcer that has been pronounced benign after a microscopic examination of its cells will seldom turn malignant.

The answer to the first question is yes—sometimes ulcer pain does coincide with cancer. The concern involves the gastric type only. Between 5 and 15 per cent of what are first thought to be simply the pains of stomach ulcers turn out to be signals of malignancy. In most such cases both ulcers and cancer are present. Generally the ulcer is not the cause of the cancer. Rather the cancer precedes and in effect creates the ulcer; the malignancy weakens the stomach wall, providing an opening for gastric juice to do its damage. Because of the possibility that an underlying cancer may be dismissed as ulcer pain, the diagnosis of a possible gastric ulcer must be pursued with the greatest thoroughness and exactitude. If the ulcer proves to be unaccompanied by any cancerous growth, there is less to worry about.

Even as doctors try to understand ulcer pains, they continue to ponder the riddle of why certain people acquire the ailment while others, who might seem likely victims, do not. No one has ever yet figured out why the disease has changed from being an affliction of well-off young women to one involving mostly males. Not only are many more men stricken than women, but different types of males seem to develop the two different types of ulcer.

Duodenal ulcer is predominantly a disease of youth or middle age, often first experienced by men in their twenties if not in their teens. Dr. Eisenberg told of a 17-year-old honor student who suddenly doubled up with excruciating abdominal pain from a duodenal ulcer while he was preparing for an examination.

Gastric ulcer, by contrast, has increasingly become a disease of older people. One typical sufferer cited by Dr. Eisenberg was a 62-year-old auto assembly-line worker of moderate habits who had rarely missed a day of work all his life; despondent over the recent death of his wife, he began feeling pain in his upper abdomen. He had developed an ulcer on the wall of his stomach.

Myth of the executive disease

It is significant that the duodenal ulcer victim Dr. Eisenberg described was an honor student, while the man who had the gastric ulcer was a factory hand. For no known reason, duodenal ulcers seem to hit high-anxiety personalities who are engaged in intellectually demanding activities, while stomach ulcers are associated with those in jobs that are physically taxing. Furthermore, the duodenal version tends to strike people who are in higher social and economic groups, and the gastric afflicts those less elevated in social standing. But the myth that ulcers are the exclusive preserve of top-ranking executives is just that—a myth. Statistics show that the disease is rife among people living in urban ghettos. One study in Pittsburgh found that factory foremen had higher ulcer rates than the executives who were their bosses. And in parts of Africa, India and the Orient, farmers and unskilled laborers proved to be riddled with the disease.

As recently as 1977, it was reported that men outnumbered women tenfold as victims of duodenal ulcers; with gastric ulcer the ratio was three to one. By 1980 the duodenal ratio had become just two to one, while the gastric rate was even between the sexes.

One reason for the male-female discrepancy may be hormonal, for events involving changes in sex hormones are known to influence ulcers. Pregnancy, for example, retards the development of ulcers; many a woman has seen her symptoms disappear after she has conceived, only to find them reappearing soon after the baby arrives. Similarly, the onset of menopause appears to increase a woman's vulnerability, and elderly men and women tend to acquire ulcers at about the same rate.

Female hormones may thus exert a protective influence on digestive tissues. This theory has not been translated into a preventive or therapeutic course because artificially administered hormones can have serious side effects—and other, safer drugs are very effective.

The recent change in male-female ulcer statistics may reflect the changing status of women. The shifting ratios of ulcer vulnerability could indicate shifts in stresses endured by the sexes—but the possible influence of psychological effects only introduces new riddles. There can be no agreement as to which sex undergoes more stress at what times. Nor is there any consensus on the importance of stress as a cause of ulcers.

Debate on this point has raged for decades, with no foreseeable conclusion. There is no doubting that stress and anxiety may be key contributing factors. The tendency of the vagus nerves, under the stimulus of emotional tumult or even of the mounting strains of daily existence, to keep the stomach churning with acid is a physiological reality.

Just how rapidly such influences can work was pointedly demonstrated in 1946 by a Chicago physician, Dr. James Clarke (in an experiment that today might be challenged on ethical grounds). During the hospital stay of a duodenal ulcer patient who happened to fancy himself an expert chess player, Dr. Clarke challenged the man to a bedside match. By means of a tube inserted into the patient's stomach, his acid output was being measured. Unbeknownst to the patient, Dr. Clarke was himself a skilled player, and he defeated the ulcer sufferer. As the game progressed and the patient's anguish increased, so did the volume of acid being collected. Hours after the match had concluded, the man's stomach was still pouring out excess acid. Not until the next day did the secretions return to normal.

The documented influence of the mind has been investigated by researchers conducting countless studies in attempts to link the effect to environment, jobs, rank and social status. The research demonstrates that all these factors can be important in ulcers—but not consistently. Their roles continue to confuse the experts.

Certain occupations seem to be particularly stressful, and thus provide a likely climate for ulcers. Most research supports this view without proving a cause-and-effect relationship. Studies have shown that surgeons, military pilots, taxi and bus drivers, train engineers, airline personnel and stockbrokers are among those most prone to ulcers. The world of high finance, in fact, has been so notorious for its tensions that ulcers were long referred to as the ''Wall Street disease''; Dr. Andrew C. Ivy once remarked that he could scarcely hope to produce ulcers in dogs until he could get them to worrying about the stock market.

More difficult to assess is the question of whether a person's rank in an organization can determine the potential for ulcer. The assumption persists that high-level business executives are more prone to ulcers than blue-collar workers or agricultural laborers.

In a now-famous experiment that was conducted in 1958, researchers at Walter Reed Army Institute of Research, in Washington, D.C., subjected two pairs of monkeys to repeated electric shocks that would be signaled in advance by a flashing red light. One monkey in each pair could prevent the shock for both itself and its partner if it would press on a bar whenever the light went on. If it failed to act in time, both monkeys would get shocked. In each pair the decision-making animal, who became known as the executive monkey, developed duodenal ulcer disease; the control monkeys did not. Additional tests showed that the timing of the stress was crucial. Ulcers developed only if the stress was intermittent—turning the animal's system on and off according to a particular schedule.

The dangerous stress: fast decisions

The Army research seems to confirm many of the investigations of occupations: The kinds of jobs that often require sudden decisions having crucial consequences are the occupations most associated with ulcers. Among the most vital decisions are those made by air-traffic controllers. Their job consists largely of decision-making—a life-or-death judgment every few minutes—and their susceptibility to ulcers and other stress-related ailments has been extensively studied. The results seem to indicate a connection, although their ultimate meaning is disputed.

In one study, Dr. R. R. Grayson, President of the Academy of Air Traffic Control Medicine, examined 111 control-

lers in the Chicago area over a year's period. He found that more than three fourths of them were suffering from ulcer symptoms, and almost a third turned out to be afflicted with previously undiagnosed and untreated duodenal ulcers. Dr. Grayson concluded that his subjects' ulcers were ''a stress-related disease, the stress being occupational.''

A decade after Dr. Grayson's report, a much more extensive investigation—covering 416 controllers selected at random from among 898 in five states—was completed by Dr. Robert M. Rose and his colleagues at Boston University. This study, too, found an unusually high incidence of ulcers among controllers: Although the researchers did not specifically blame the stress of the job, they noted a prevalence of ulcers ''almost triple the national rate.'' On the other hand, a statement that was published in the journal *Aviation, Space and Environmental Medicine* said: ''In general the stress of ATC work is no greater than could be expected for 'normal' populations.''

The disagreement over the impact of the air-traffic controllers' work is complicated by the fact that external stress alone does not consistently cause ulcers. Many competitive types thrive under stress. Commented one gastroenterologist, ''That some of us have ulcers is curious. The wonder is that we all don't.''

The answer to this puzzle may lie not with a person's circumstances—job, status or sex—but with something deeper and more innate: personality. Perhaps some people are burdened with intrinsically ulcer-prone psyches, and are thus more likely to succumb to the ailment under stress. Psychiatrist Franz Alexander of the University of Illinois maintained that individuals who were caught in the grip of a profound and insoluble inner conflict were the most likely ulcer victims. They might fancy themselves strong and independent, for example, while deep down they crave protection; or they may feel a need to be aggressive but are never able to really behave that way. ''Ulcer patients,'' wrote Yale's Dr. Howard Spiro, ''as a group tend to repress their emotions or tend at least to inhibit outward expressions of strongly held feelings.''

Some physicians and psychiatrists continue to hold this

view, but it also has been challenged. Said Dr. Robert Kemp, "I have never found any evidence among my patients suggesting that there is such a thing as an 'ulcer personality.' They seem to me to be a complete cross-section of all types of emotional make-up." In one study, conducted by doctors from the Walter Reed Army Institute of Research and the University of Pittsburgh, 2,073 draftees were given exhaustive tests to determine which of them exhibited personality characteristics like those originally described by Dr. Alexander, and were also heavy secretors of gastric juices. All of the subjects who eventually developed ulcers did secrete excessive amounts of digestive fluids, and also suffered from internal conflicts. What was confusing was the fact that just as many of the subjects were people who had the same traits but got no ulcers at all.

Medical researchers are on somewhat firmer ground when probing the role of heredity in causing ulcers. While your ancestors cannot bequeath you an ulcer, they do seem able to predispose you for or against one. Anyone who has an ulcerous brother or sister is two and a half times more likely to get one than someone whose family history is free of the ailment. If one identical twin gets an ulcer, reported Dr. Jon Isenberg of the University of California at San Diego, the other twin—whether an executive or a ditchdigger—has almost a 50 per cent chance of acquiring one too.

Some families are extraordinarily victimized. An extreme case is that of a Topeka, Kansas, machinist who suffered from a duodenal ulcer: His grandfather, father, two of his five brothers, one of his two sisters, and a niece and nephew all were ulcer patients as well.

Blood type, which is inherited, has also been correlated with the frequency and variety of ulcer: Type O blood is seen more commonly in patients with duodenal ulcer, while those with gastric ulcer more often have type A blood. Individuals with type O blood, furthermore, are 35 per cent more likely to develop duodenal ulcers than people with other blood types. Some doctors believe that type O blood may be deficient in certain defensive chemicals, called antigens, that might protect against ulcer formation.

None of these genetic traits foretells who will get an ul-

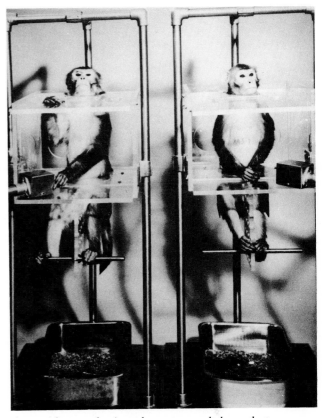

As part of a now-classic study on stress and ulcers, the two monkeys above were given electric shocks every 20 seconds for six hours, then given no shocks for six hours of rest. The animal on the left—the "executive monkey"—could make a decision to prevent the shocks by pressing a lever (in its left paw). The other monkey's lever was useless. After 23 days, the executive monkey was found to be suffering from a duodenal ulcer. At first experimenters thought that the stress of decision-making caused the ulcer. But later experiments suggested that the six-hour cycle corresponded to peak levels of digestive activity in the monkey, and that it—and not the stress alone—was to blame.

cer—only who might. But a group effort unearthed a measurable characteristic that may help predict who is especially susceptible to duodenal ulcers. The research, which enlisted seven scientists at four institutions in Texas and California, found a possible genetic marker in families with a strong history of duodenal ulcer. Scientists had long suspected that the chemical pepsinogen—from which the potently corrosive pepsin is derived—may be just such a marker. Now it was learned that among those test subjects with high blood levels of pepsinogen, 40 per cent had contracted duodenal ulcers, while those with normal levels had no ulcers. What is more, the tendency to excess pepsinogen is inherited, traveling from generation to generation according to a predictable genetic pattern. Half the offspring, both male and female, of parents with abnormally high pepsinogen levels were found to possess a similar overabundance.

From this and other studies, researchers calculated that anyone carrying the pepsinogen marker is five times as likely to develop a duodenal ulcer as someone who does not have it. Because the marker is easily identifiable, the day may come when a simple blood test will be used to identify such potential ulcer sufferers.

Does diet make a difference?

Although all manner of forces, many of them beyond individual control, might predispose a person toward ulcers—heredity, emotional make-up, sex, type of job—the trigger might well be something that each person can control. Food, drink, drugs and personal habits have received their full share of examination. Some likely villains have been named, but with few exceptions they have been exonerated.

Faulty diet, of course, has long been assumed to be a culprit. But now it appears to be immaterial to the development of ulcers (although not necessarily to treatment after they develop). Some researchers have blamed ulcers on protein deficiency or lack of vitamins; there is no proof. Malnutrition has likewise been discounted. Lack of fiber, while it may give rise to other gastrointestinal ills *(Chapters 2 and 3),* seems to have no causative effect on ulcers—nor does a high-fiber diet help prevent them, much as it may benefit

the intestines. Although foodstuffs high in acid, like many fruit juices, might seem to foment ulcers, such is not the case: Grapefruit juice, notable for its sharply acid taste, will not stimulate the flow of hydrochloric acid and pepsin any faster than other beverages. Hot and spicy foods, such as garlic and curry powder, seem to inflict little harm on the stomach's mucosa.

However, Dr. G. P. Crean of the Gastrointestinal Centre of Southern General Hospital in Glasgow noted, "It is difficult not to believe that dietary factors might exert some influence." He added, "That particular items of the diet may cause severe mucosal damage in the upper alimentary canal is suggested by the observations that a variety of Japanese pickle" may badly inflame the stomach lining.

Among the foods often blamed for initiating ulcers is coffee. The stimulating compound it contains—caffeine, which is also an important ingredient of tea, cocoa, chocolate and cola drinks—was found by many separate investigations to increase acidity in the stomach, although a study done by the U.S. Navy came up with contradictory evidence. But the caffeine in coffee is apparently not the only acidifier. Tests of decaffeinated coffee found it to increase acidity only slightly less than regular coffee—and much more than purified caffeine alone. Even the coffee-like beverage made by brewing grounds of roasted grain—grain that is free of any caffeine and totally unrelated to the coffee bean—has a marked effect on stomach acidity. Researchers speculate that some effect of the roasting process, whatever it is applied to, may introduce acidifying substances other than caffeine.

Whether these acidifying effects—from the caffeine or something else in coffee—can cause ulcers remains in some dispute. Some authorities believe they can. As part of a massive survey of the health of nearly 40,000 Harvard and University of Pennsylvania alumni, Ralph S. Paffenbarger Jr. of the California State Department of Health found with his associates that among those alumni who drank two or more cups of coffee per day, ulcer incidence was nearly 80 per cent higher than among the subjects who drank no coffee.

Less equivocal than the case against coffee is the indictment of some substances that are consumed but are not food-

Out of the mouths of frogs, babes

A rare little frog, *Rheobatrachus silus,* found in 1973 in the mountains of southeast Queensland, Australia, may lead to new discoveries about digestion. Unlike any other animal, this frog nurtures young in the stomach and gives birth through the mouth *(below).* An explanation of how the digestive system is temporarily shut off, stopping the flow of acids and allowing the young to develop unharmed, may lead to help for people with gastrointestinal problems, particularly ulcer sufferers.

The female lays eggs and the male fertilizes them in the normal fashion. Next, however, the female gobbles up the eggs and, instantly, her digestive system shuts down. Acids stop flowing, stomach and intestinal muscles stop working, and food recently eaten stays in the intestine. For eight weeks the frog's digestive system remains dormant while the eggs turn into tadpoles and then into little frogs. When the babies are ready for birth, the mother opens her throat and the young, propelled by the mother, pop into the mouth and then jump out into the world.

Research by Australian zoologist Michael J. Tyler and his colleagues suggests that this reproductive scheme is made possible by a chemical, perhaps a hormone, that halts the mother's digestion. It is released, they speculate, not by the mother but—somewhat surprisingly—by the eggs, tadpoles and juvenile frogs.

*A fully formed baby frog wriggles out of its mother's mouth.
The female of this unique Australian species eats nothing while the
brood develops inside her stomach but resumes feeding soon
after giving birth. Within eight days her stomach, grossly swollen
during the brooding phase, returns to almost normal size.*

stuffs: cigarette smoke, alcohol and certain other drugs, notably aspirin. Cigarette smoking not only is implicated in heart and lung diseases but also seems to stimulate ulcers. According to the Center for Ulcer Research and Education in California, heavy smokers of cigarettes double their risk of contracting ulcers; and when they do have the disease, their ulcers heal more slowly. Said Dr. Nathaniel Cohen, ''I would urge giving up smoking for any number of reasons, not excluding ulcers.'' Still, some physicians believe that the stress involved in kicking the cigarette habit may be more harmful to certain people—and thus more ulcer-inducing—than continuing to smoke.

As for alcohol, it surely stimulates acid flow. Furthermore, direct application of alcohol to the stomach lining, as by the swift quaffing of a martini, can cause cell-wall inflammation; although the cells will normally repair themselves, repeated injury may break down the cells and allow acid to leak back into the stomach wall. Alcohol also intensifies the deleterious effects of caffeine, and is a direct cause of gastritis—inflammation of the stomach lining—which can lead to ulceration. All such damage depends on heavy use of alcohol, however. By itself, in moderation, liquor cannot be said to induce ulcers.

In the case of aspirin, doctors have no reluctance to assign blame. ''Aspirin is a marvelous drug,'' said Dr. Grossman, ''and should be used where there are medical reasons.'' But he also said, ''There is no question that excessive use of aspirin is a cause of gastric ulcer.'' In some

Beaming lasers at bleeding ulcers

The remarkable crystal wand pictured below provides a new way to locate and stanch bleeding ulcers in a single stroke. Called a wave guide, the threadlike quartz filament creates a flexible pathway that enables laser beams to bend around corners. The device is fed into the stomach through a fiber-optic endoscope—a diagnostic tool that, when snaked through the throat and down to the stomach, provides a kind of gastric peephole. Once the wave guide has been aimed at the hemorrhaging ulcer, a laser beam is fired down the filament to create a spot of intense heat that sears the ruptured vessel shut.

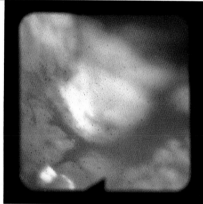

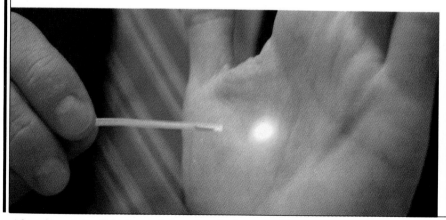

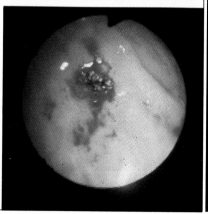

A harmless aiming light, beamed through a plastic-encased laser wave guide, brightens a physician's palm. In the stomach, the light helps pinpoint a bleeding ulcer before the laser is fired.

Blood wells up around an untreatd whitish gastric ulcer (top). A similar ulcer (bottom), photographed from slightly farther away after laser treatment, is covered with a charred scab, which will give way in a few weeks to healthy tissue.

people, said Dr. Cohen, even small amounts are harmful.

Many different kinds of remedies contain aspirin as one ingredient—a fact that sometimes is revealed only in small type on the label. Anyone who suspects he is ulcer-prone should check labels carefully.

Signs that mean life or death

Given the fog surrounding most of the presumed causes of ulcers, foolproof techniques for avoiding the disease do not exist. But because ulcers seem to result from a number of factors working together, anyone who qualifies as a potential victim on one account ought to be careful on some of the others. If your family has a history of ulcer disease, if your way of life is particularly stressful or your job especially high-pressured, prudence suggests that you avoid aspirin and be sparing in your use of alcohol, tobacco and caffeine. In short, practice moderation.

Beyond that, the best prevention of severe ulcer disease is early diagnosis and prompt treatment. It pays anyone—even those without telltale proclivities—to know about ulcer's symptoms; if any show up, a doctor should be consulted promptly. The signs are easy to recognize once you understand them, but they are very easily missed or dismissed as unimportant by the uninformed. Many a victim has learned of his ailment only after he was saved from bleeding to death in a hospital emergency room.

The primary symptom is pain. Most stomach pain, of course, is merely indigestion. But if agonizing pain strikes suddenly and without warning, prompt professional care is essential. When a gnawing sensation in the pit of the stomach—just below the rib cage—comes on suddenly and is acute, it could signal a perforation; in such cases a previously undetected ulcer has eaten all the way through the wall of the stomach or of the duodenum. Such pain could also be a symptom of appendicitis, however, or of gallstones. In any case, consult a doctor immediately.

More often the pain will be milder, at least at first, and it makes sense to observe its behavior for a week or two. If the cause is ulcer, it will identify itself as something more than indigestion by its more-or-less regular appearance. Among most people it tends to follow eating, right on schedule. If it occurs shortly after dessert and is felt to the left of the midline, just below the ribs, the implication is a gastric ulcer. If an hour or two after eating the pain comes on just to the right of the midline, it may mean a duodenal ulcer. The pains of duodenal ulcer may mimic the symptoms of other ailments, such as backache or liver disorder; its clocklike relationship to eating will help differentiate it. But even when the signs point unmistakably to an ulcer, do not attempt to treat yourself. Above all, take no medicine, for many nonprescription drugs can worsen an ulcer.

One symptom that calls for prompt attention from a doctor is bleeding. If it manifests itself in vomit, which will be red or coffee-colored—the blood having chemically reacted with acid—get attention immediately. If small amounts show up in the stool, there is no immediate emergency, although medical care should not be unduly delayed.

Blood in the stool is not obvious. If it comes from an ulcer, it will usually produce a black or tarry effect, the blood having been changed chemically during its transit through the intestines. If red blood appears in the toilet, the cause is probably not ulcers. Such reddened stools are likely the product of hemorrhoids or a bowel disease *(Chapter 5),* for blood that has not changed color must come from a point close to the end of the large intestine, not from the upper part of the digestive tract affected by gastric or duodenal ulcers. Any indication of bleeding anywhere in the digestive tract must not be ignored, however, because it can be a signal of a dangerous disease, including cancer.

Often the doctor can deliver a swift and accurate diagnosis of ulcers just by taking a medical history—learning when the pain showed up and what other abdominal troubles may have preceded it, then asking about family history, job situation, living habits and general state of mind. He will also conduct a physical examination, feeling for tenderness or other obvious abnormalities in the abdomen. Almost certainly he will call for an "upper GI series," a set of special X-rays that show the stomach and duodenum. These organs, made of lightweight, soft tissue, are ordinarily transparent as glass to X-rays, but they can be made visible if the patient drinks what radiologists like to call a barium milkshake—a raspberry- or chocolate-flavored concoction containing the element barium. Barium is very dense and is thus opaque to X-rays; when the liquid enters the stomach and duodenum, it outlines their walls in the X-ray, revealing any nicks or holes that indicate an ulcer. More than 90 per cent of all ulcers show up on this kind of X-ray; a skilled radiologist can even detect from the outline of the digestive wall signs of scarring left by previous ulcers.

If the evidence points to a duodenal ulcer, no other tests will likely be needed. A gastric ulcer, however, requires further and sometimes protracted examination to rule out the possibility of cancer. Not long ago, this essential step might have required surgery. Today it does not, but is almost a routine procedure, thanks to the fiber-optic endoscope (*pages 92-95*). This tool is an intricate, flexible tube that the physician can feed through the throat into the stomach to make an inch-by-inch survey of the digestive wall, photo-

graphing specific areas of concern and even, by means of diminutive pincers or wire lassoes, obtaining tissue samples to analyze for malignancy. An ulcer will appear as a tiny—or not so tiny—crater with a white base and sometimes a red rim. Other tests can also be made, such as withdrawing stomach fluid through a tube and analyzing it for acid content. But the endoscope, sometimes used repeatedly, has made most of these procedures unnecessary.

Two revolutionary drugs

Assuming that the tests reveal no malignancies, and no other complications, the doctor can prescribe immediate treatment. The prescription has been revolutionized by new techniques: Until the 1970s, many ulcer patients would routinely be clapped into the hospital, placed on a schedule of total rest and subjected to a severely restricted diet; surgery (*Chapter 6*) was common.

Most of the rigors of ulcer treatment have been eliminated

The revealing GI series

Among the most common tests for gastrointestinal disorders is the barium X-ray. X-rays ordinarily pass through the soft tissues of the digestive tract; but when the organs are filled with barium sulfate, which blocks the rays, the organs come clearly into view on film.

The addition of a second chemical compound can make the standard barium test even more effective for diagnosing such ailments as cancer and ulcers. In a so-called double-contrast barium study (*right*), the patient swallows a compound that produces carbon dioxide gas, then downs a chalky ''milkshake'' containing barium sulfate. The gas expands the organs being studied and pushes the barium into every nook and cranny of their internal walls. Since the barium contrasts on X-ray film not only with the organs it coats, but also with the gas—a double contrast—the pictures produced are more distinct.

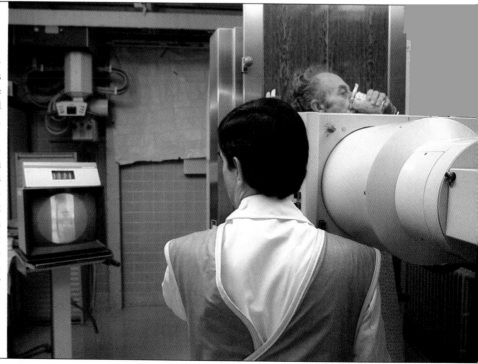

A patient (background) drinks a milkshake of strawberry-flavored barium sulfate while Dr. Paul Frank of the University of Chicago, wearing a lead apron to protect himself from X-rays, watches a monitor that shows the solution trickling down the esophagus. At intervals the doctor took still pictures of the barium as it moved through the upper digestive system.

by new drugs that make the lesions heal quickly—almost magically, it seems to doctors and patients alike. The first of these medicines to be used in the United States and Europe was cimetidine (better known in America by its American trade name, Tagamet), which simply stops the stomach from producing much acid.

Medical science has long been aware that one of the body's natural chemicals, histamine, plays a part in the secretion of stomach acid. Histamine is the same substance that causes a cold victim's nose to run, and it is what makes a bee-sting itch and swell. Just what role it takes in the process of digestion was for many years unclear, but doctors theorized that it stood between gastrin and the acid and pepsin of the gastric juices. Gastrin, in other words, was presumed to stimulate histamine, which in turn would provoke the secretion of the digestive juices. The object was to find a substance that would block the action of histamine in the stomach. The standard antihistamines did not do it: They inhibit only one type of histamine-sensitive chemical trigger in human cells, the so-called H_1 receptors. The chemical structures in stomach cells, called H_2 receptors, were impervious.

The breakthrough was finally made at the British laboratories of Smith Kline & French by a team of researchers led by Sir James Black, the physician-scientist who also helped develop propranolol, one of the drugs known as beta blockers. Propranolol prevents recurrent heart attacks and treats many other ills. Sir James began his search in 1964 and over the next six years his team worked its way through 700 compounds before synthesizing one that safely blocked the action of the H_2 receptors.

Several years before cimetidine became available, Japanese physicians began using a quite different drug, sucralfate (given various trade names in different countries, including Carafate in the United States, Sulcrate in Canada and Ulcogant in Germany). Sucralfate works by interacting with a protein substance that is secreted by the ulcer itself,

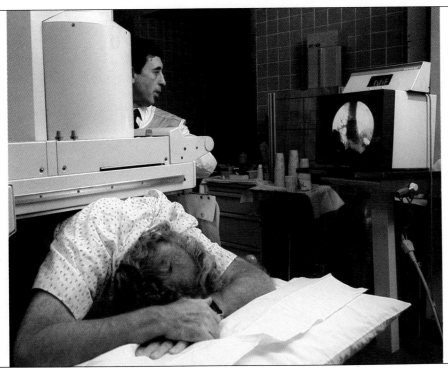

Filming his patient from overhead, Dr. Frank monitors the screen for unusual barium concentrations in the esophagus—a sign of esophagitis, caused by stomach acids backing up into the esophagus. The potentially harmful condition usually results from improper functioning of the lower esophageal sphincter, the valve located where the esophagus meets the stomach.

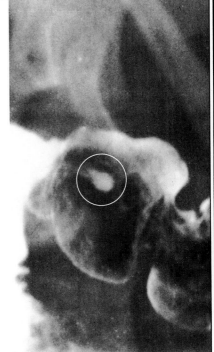

Circled in white, an ulcer in the duodenum, or upper intestine, reveals itself on a double-contrast barium X-ray. In the meantime, most of the barium compound, seen as the light area to the left, has progressed farther down into the intestine.

to form a coating over the sore that acts as an effective barrier against gastric juice. Thus protected, the ulcer can heal unmolested. A controlled test at the University of Montreal found that sucralfate promoted healing of duodenal ulcers in four weeks. Because it acts primarily as a surface ''bandage'' and is virtually not absorbed into the bloodstream, it seems unusually safe, with no known effect upon other ailments or drugs.

Thanks to such potent medicines, a hospital stay is no longer required in all ulcer cases. Patients with a gastric ulcer may be hospitalized for a week or so, as they often tend to heal faster under close supervision. Anyone with a perforated or hemorrhaging ulcer or other complication will surely be admitted. Some older people may be better off in the hospital, as will those whose home atmosphere is so tense as to aggravate their condition. Most others are thought to be better off going about their business and being treated in the doctor's office. ''If a patient is an important executive who is about to land an important contract,'' said Dr. Nathaniel Cohen, ''I know he'll be worse off being taken away from his job at this juncture.''

Rest is still considered necessary, to calm the vagus nerves. But rarely total bed rest. The standard advice is to get more sleep, and to avoid emotional turmoil. Otherwise lead a normal life. ''Everything is OK, in moderation,'' wrote Dr. G. S. Serino of Wilmington, Delaware. ''Everything, that is, but strife.''

New challenges to the milk prescription

Until the 1970s, the requirement for rest was almost always coupled with a strict bland diet. The most popular nutritional regimen was worked out by Dr. Bertram Sippy of Chicago, who in the years before World War I recommended that an ulcer be managed by means of hourly feedings of milk, cream, eggs and cooked cereal. Thus was born the perennial image of the ulcer patient dutifully sipping his milk. The theory was that bland foods, particularly milk and cream, are effective neutralizers of stomach acid. In addition, because of their high fat content, both take a relatively long time to digest, and would thus tend to remain in the stomach, coating

Drugs that heal the wound

Aside from their acid origins and characteristic gnawing pains, the two major types of ulcers—gastric and duodenal—have relatively little in common. They have different locations in the intestinal tract: Stomach, or gastric, ulcers occur most often in the gastric antrim, near the pyloric sphincter (drawing below); the duodenal type, four times more frequent, attacks the duodenum, the portion of small intestine just beyond the pyloric sphincter. Gastric ulcers, which afflict more older people, seem to result from a breakdown of the stomach's normal defenses against its own acid juices; duodenal ulcers affect people who produce too much acid and pass it on, only partially digested, to attack the lining of the duodenum.

But because both kinds of ulcer are formed by the corrosive effects of acid, both are improved—and often cured—by two new drugs that keep acid at bay. The drug cimetidine (sketch at right) thwarts the production of hydrochloric acid at its source; sucralfate (far right), which was developed in Japan, blocks acid after it is formed and shields the ulcer site against additional harm.

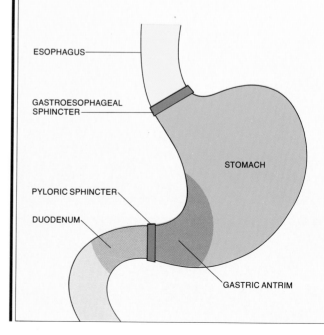

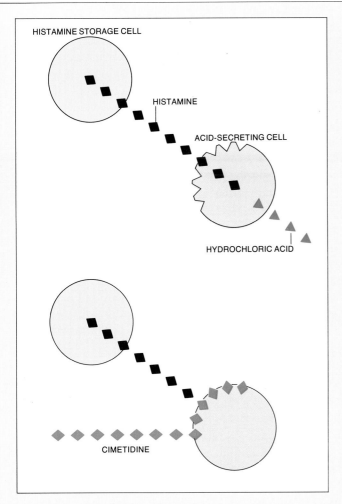

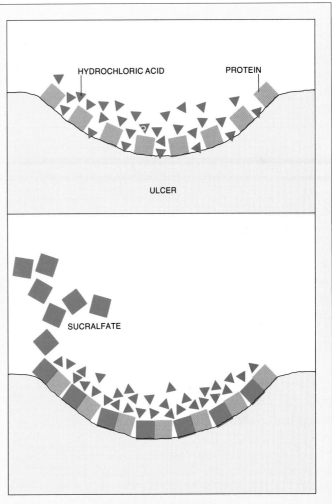

A CHEMICAL TO TURN OFF ACID FLOW

In normal digestion (top sketch), histamine, released from storage cells in the stomach lining, triggers production of hydrochloric acid by fitting into receptor sites on acid-secreting cells nearby. The lower sketch shows how cimetidine molecules wedge themselves into the receptor niches to block histamine activity.

A CHEMICAL BANDAGE TO SHUT ACID OUT

Left untreated (top sketch), the raw ulcer bed is irritated by hydrochloric acid that seeps between protein particles oozing from exposed tissue. In the second sketch, sucralfate molecules chemically combine with this protein to form a sticky coating that keeps out acid and shields the ulcer bed.

and protecting its injured lining and allowing ulcer inflammation to subside.

The Sippy Diet has been largely abandoned as a long-term treatment. For one thing, some patients found it so burdensome that their stomachs actually produced more acid out of sheer vexation. More important, milk was discovered not to reduce acid but to foment it. Milk proteins initially boost the stomach's alkalinity, but as acid level is reduced, the hormone gastrin stimulates still more acid and pepsinogen. Today most experts agree with Dr. John S. Fordtran, of the University of Texas Southwestern Medical School, that hourly milk and cream intake serves only to maintain stomach acidity. Some doctors tell their patients to stay away from milk altogether.

The harm—or benefit—conferred by milk is still debated, however. The author of a definitive text, Dr. Howard Spiro, in an article in the *Journal of Clinical Gastroenterology* entitled "Is Milk All That Bad for the Ulcer Patient?" recounted the following incident: "Not too long ago I was drinking 'Pisco Sours' in Peru with a colleague, coeditor of 'another' gastrointestinal textbook. I was enjoying the conversation, when my drinking companion said to me, 'Howie, I've got terrible heartburn—I've got to get a glass of milk!' Looking at him in disbelief, I exclaimed, '. . . What? You're the guy who says that milk is no good! Your book teaches that milk makes ulcer patients worse because it raises the serum calcium and increases gastric acid—and you want a glass of milk! For heartburn!'' He sighed, and replied, 'Yes, I know all that. But milk makes me feel so good!' "

Dr. Spiro pointed out that "the legendary surcease of milk may in fact have a physiological basis," citing the work of scientists at The Wellcome Research Laboratories in North Carolina, who discovered that milk can dull pain. A compound formed in the body with a fragment of a milk protein acts like morphine, they found, blocking the transmission of pain signals by nerve cells. It fits into the same nerve receptors that are ordinarily filled by opiates, in effect turning off the nerve cell.

Most doctors agree with Dr. Spiro's practical view of the confusion over diet. They continue to prescribe a bland diet, rich in milk and creamy foods, for a patient recovering from a severe ulcer attack—it makes the patient feel better. But once the crisis is past, most dietary requirements are lifted. "If something seems to agree with you," said gastroenterologist Nathaniel Cohen, "go ahead and eat it." Because food itself tends to blunt the onslaught of gastric acid, patients are often advised to eat more often—perhaps five or six times a day. Suggested Dr. Spiro, "Only three rules are pertinent: (1) Eat between your meals. (2) Do not get hungry. (3) Whatever does not hurt you is permissible."

Just a few substances are generally ruled out. The American Dietetic Association lists as items to be avoided because they irritate the stomach: black pepper, chili powder, caffeine, coffee, tea, cocoa and alcohol. Some doctors do not entirely proscribe alcohol and tobacco. "If you enjoy a relaxing drink, don't give it up," said Dr. G. S. Serino. "Drink taken in moderation adds a little relaxation and conviviality to life." And Dr. Spiro wrote, "Any harm to the patient's stomach from the amount of nicotine derived from smoking is more than outweighed by the pleasure and feeling of well-being that cigarettes are said to induce in those already addicted to them."

As the ulcer patient experiments with various foodstuffs, his physician will surely prescribe a course of medication that must be observed faithfully. One of gastroenterology's most durable slogans is, "No acid, no ulcer," and even victims of gastric ulcer, an ailment accompanied by normal acid levels, will probably be given an antacid. Perhaps it will be one of the liquid products that has proved to bring the fewest side effects *(Chapter 2)*.

Together with nonprescription antacids, many doctors continue to prescribe the use of cimetidine because this drug prevents the recurrence of ulcers—indeed, it is widely prescribed for people who are suffering digestive upsets but have only incipient ulcers. It works particularly well when taken in conjuction with an antacid, the cimetidine reducing the gastric flow and the antacid neutralizing any residual acidity. The favored time for ingesting both is in the evening, just before retiring, to minimize the likelihood of a 2 a.m. ulcer pang. Like any systemic drug, cimetidine has been known to cause a few side effects—minor dizziness and muscular pain, among others.

Miraculous though the new medicines seem, they remedy only physiological effects, not the underlying causes of ulcers. In view of the probable role that stress and emotion play in generating and perpetuating this ailment, relief may depend on uncovering the particular factors—physical or psychological—that brought the illness on, so that these conditions can be corrected. Some patients will be able to do this simply by talking with the doctor. Others may benefit from psychiatric counseling.

The principal therapy in any case remains a judicious combination of drugs, diet and rest. Such a regimen relies on the body's own superlative powers of recuperation. Doctors simply isolate and protect the ulcer so that the tissue can heal itself. Once shielded from the corrosive gastric juices, the ulcer crater should fill in, scarring over in a few weeks. Under most circumstances the stomach or duodenum should return to normal, with only a slight scar to indicate the trouble that has occurred.

How to live with a chronic ulcer

But along with "No acid, no ulcer," one of medicine's abiding rules is, "Once an ulcer, always an ulcer." For even though the sore has healed, the ulcer propensity has not really been removed. No true cure for ulcer disease is known. The ulcer itself has been a manifestation of an imbalance that medical science to date does not know how to correct. The ulcer-prone person continues to be vulnerable, and if he takes no precautions his ulcer may recur in the same spot, or elsewhere, a few months or a few years later. One Veterans Administration study showed that 40 per cent of gastric ulcers are likely to recur within two years.

Most patients learn to accept the recurrences and modify their habits when the pain returns. A 42-year-old Hawaiian orchid grower found he could lead a perfectly conventional existence between ulcer attacks. But when one arrived, he promptly put into effect a program he and his doctor had learned would work. He cut down on his smoking, limited his coffee to one or two cups a day and eliminated alcohol entirely; meanwhile he started to take medicine again. After three weeks or so, the pain usually disappeared, and he returned to his former habits.

Some ulcer victims are not so lucky. Their ulcer for one reason or another may get worse, or even in the early stages it may threaten severe complications. In such cases the victim must consider surgery.

A doctor may recommend surgery if the ulcer has perforated. Almost invariably this necessitates swift action to seal the stomach or intestinal wall so as to prevent the dangerous infection of peritonitis. Hemorrhaging may also call for surgery. Bleeding can generally be controlled medically, but should an ulcer open up a large blood vessel, an operation is usually mandatory. A third reason is intestinal obstruction: Sometimes scarring, swelling or inflammation from repeated ulcer attacks can result in partial or complete blockage of the afflicted region, and a surgeon must then remove the block. A final condition is intractability: A constantly recurring ulcer, or a worsening one, may become such a burden that only surgery can restore the patient to a normal life.

Most ulcer patients learn to live with their vexing situation, however. Some seem to be able to escape from torment entirely. One unusual case recounted by Dr. Spiro concerned a 40-year-old man who was a practicing physician.

"For many years," wrote Dr. Spiro, "he had been a curmudgeon, chasing some patients from his office, giving salty advice to others, and generally raising cain." Then one day he decided that he would change. "He would become sweetness and light," recalled Dr. Spiro, "and, life being short, he would be kind to everyone regardless of how he really felt. Exactly on the day that he made that decision, he first suffered severe ulcer pain." The pain grew steadily worse, finally sending the reformed curmudgeon to the hospital, where he was found to have a deeply penetrating duodenal ulcer that would probably require surgery. Faced with that prognosis, "the patient recognized exactly what had happened and what his cure should be. Soon after he resumed his normal caustic and dour exterior, the ulcer for which operation seemed so inevitable stopped paining him and it has not recurred."

A new tool kit for the gastroenterologist

The only certain method for finding out what ails the digestive system has been exploratory surgery—operations that may prove more dangerous than the disorders they reveal. But now this risk can generally be avoided with the gastroenterologist's new tool kit: a variety of ingenious devices that—without requiring the abdomen to be cut open—provide detailed views of the inside of all 30 feet of the digestive tract. Some of these instruments can even extract samples for laboratory analysis or make minor repairs. The new techniques are standard at well-equipped hospitals such as The University of Chicago Medical Center, where the photographs on these pages were taken.

Perhaps the most revolutionary of the new tools is the family of viewing devices called endoscopes—long, flexible instruments that look inside the tract and work on it. One type, a panendoscope *(right)* is inserted through the mouth to view the esophagus, stomach and even the duodenum, enabling doctors to check for ulcers or cancers.

Although the endoscopes open almost the entire gastrointestinal tract to view, some problems are best diagnosed by other techniques. To pinpoint bleeding in such organs as the liver and pancreas, specialists rely on angiography *(pages 96-97):* A dye is injected into an artery, and the marked blood is revealed by X-rays as it courses through the vessels of the organs. To locate tumors, high-frequency sound is bounced off the organs to illustrate their faults on a screen *(pages 98-99).*

And there are remarkable new tests for less serious disorders as well. The efficiency of a slow-working stomach can now be measured by tracking the progress of a radioactive meal through the system with a nuclear scanner. Severe and unrelenting heartburn is diagnosed by a meter that senses acid backing up into the esophagus. And a common trouble—inability to digest milk—is confirmed by a simple, painless breath test.

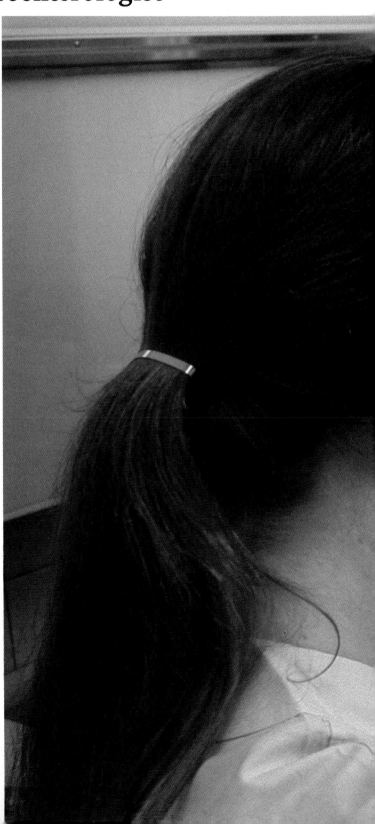

Viewing their work through the twin eyepieces of an endoscope, a gastroenterologist and a nurse prepare to remove a tissue sample from an inflamed esophagus. With his right hand the doctor maneuvers the viewing lens and surgical attachments inside the patient; with his left he directs the flow of cleansing air and water.

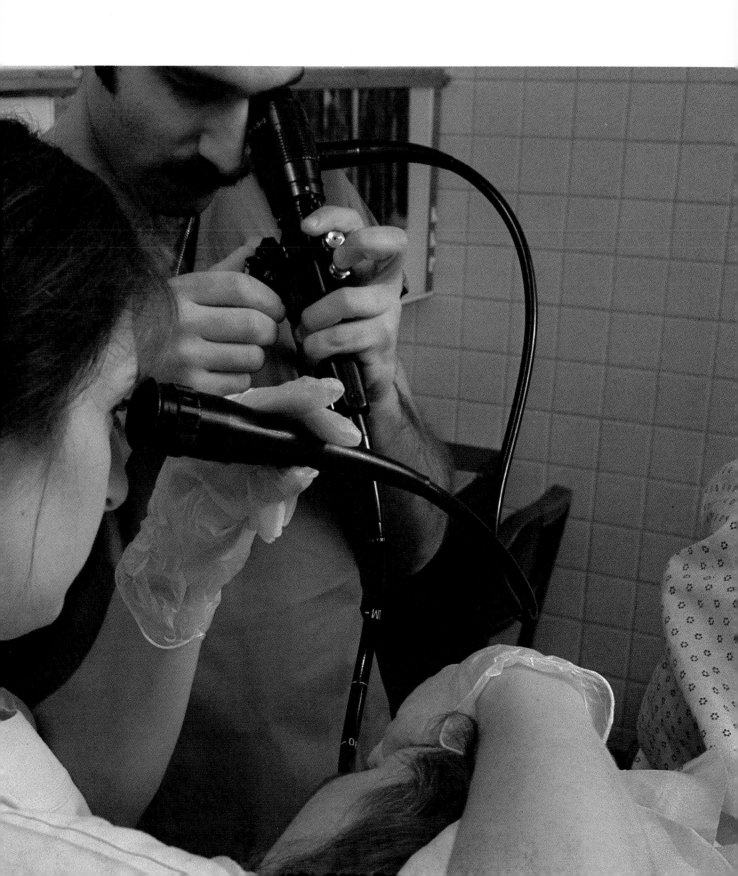

Endoscopes to look and work inside the body

A number of different types of endoscopes give the specialist access to various parts of the body. The longest view is provided by the colonoscope *(below),* which inspects, samples and repairs the large intestine. In two thirds of the colonoscopies performed, the device travels the full length, about 5 feet.

Like all endoscopes, the colonoscope is a long, flexible tube containing as many as 50,000 strands of glass fibers, each .0004 inch in diameter. It serves as a light pipe, the glass fibers bouncing light inside to transmit clear images around curves and corners from a lens at one end of the instrument to an eyepiece or camera at the other end. Most endoscopes have four-way controls that swiv-el the lens at the probing end so that the tool can look sideways, up and down. Channels in the tube permit the physician to slip in special instruments—forceps, electrical cauterizers, even lasers—for work inside the patient. Through another channel, a jet of air or water cleans the lens or tissue for better viewing.

For the colonoscope, surgical attachments are the most important accessories. If the physician sees a polyp, he can remove it with a surgical snare *(opposite, bottom),* slipped through a channel. As many as five such growths can be removed in a single session. Another attachment, a biopsy forceps, extracts tissue samples so that they can be tested for malignancy.

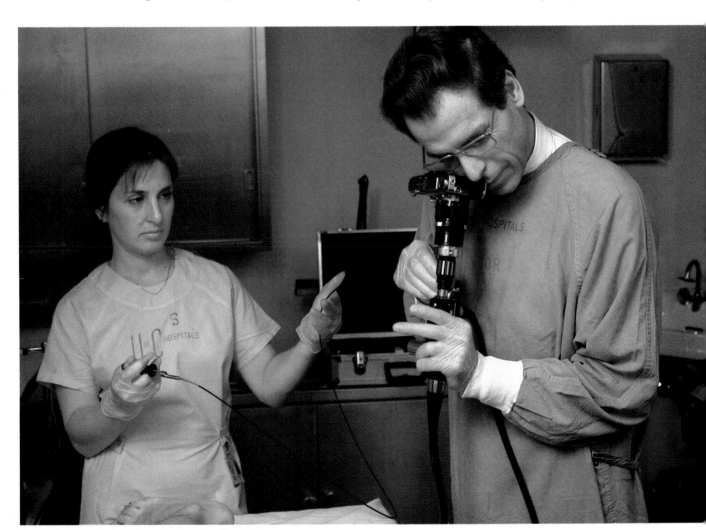

Examining a patient's inflamed colon, Dr. Gerald Rogers photographs the inside of the intestine with a camera fitted to a colonoscope. At left a nurse readies a cable that will be inserted through a colonoscope channel to operate the forceps shown opposite. The lever in her right hand will control the forceps to snip a tissue sample for testing.

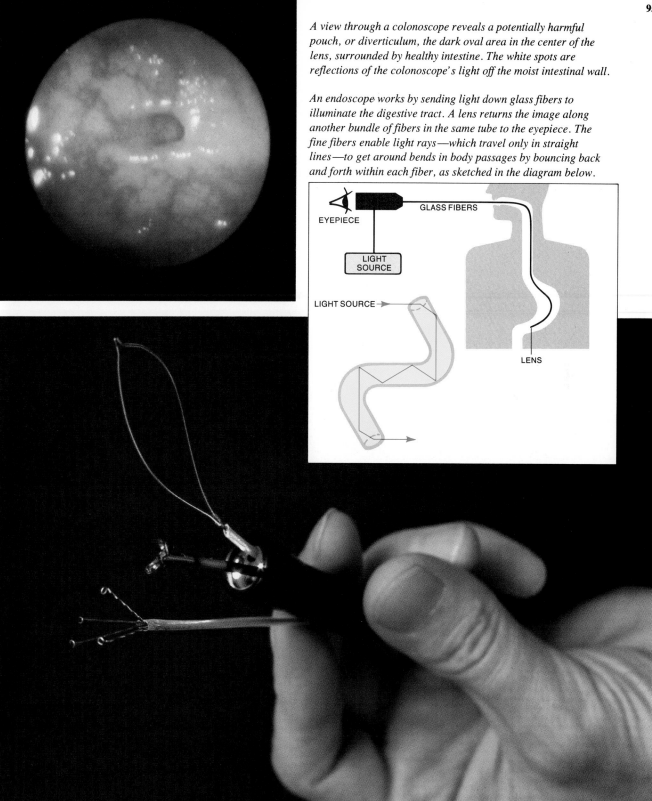

A view through a colonoscope reveals a potentially harmful pouch, or diverticulum, the dark oval area in the center of the lens, surrounded by healthy intestine. The white spots are reflections of the colonoscope's light off the moist intestinal wall.

An endoscope works by sending light down glass fibers to illuminate the digestive tract. A lens returns the image along another bundle of fibers in the same tube to the eyepiece. The fine fibers enable light rays—which travel only in straight lines—to get around bends in body passages by bouncing back and forth within each fiber, as sketched in the diagram below.

EYEPIECE

GLASS FIBERS

LIGHT SOURCE

LIGHT SOURCE →

LENS

Sheathed in thin tubing, three wire instruments—top to bottom, a snare, biopsy forceps and grasping forceps—protrude from channels of a colonoscope. The attachments, retracted inside the tubing, are threaded through the channels and then pushed out. The snare and grasping forceps cut off polyps; the jaws of the biopsy forceps snip tissue samples.

Dyes to spot blood-vessel damage

A detailed view of an organ's circulatory system can make certain digestive disorders—particularly ulcer, tumors and the inflammation of diverticulitis—much easier to diagnose and treat. A clear picture showing irregular branching of blood vessels can reveal a malignancy that has tapped into an organ's blood supply. Finding dangerous internal hemorrhaging, often so slow or intermittent that it cannot be detected by an endoscope, could indicate a serious ailment; it might even save a life.

To map an organ's circulatory system, gastroenterologists have adopted a procedure often used to diagnose heart ailments: angiography. In this technique a tube is inserted into a blood vessel and painstakingly threaded through the circulatory system to the organ. An opaque dye is then injected into the tube, and the course of the dye, as it passes through blood vessels or leaks out of them, is followed by taking successive X-ray pictures. Bleeding shows up as light splotches.

The patient, who is sedated but still conscious, experiences little discomfort during the procedure. Once the catheter is in position, the dye is injected. Immediately the patient feels a transient rush of heat. It is during these first 30 seconds after injection that the X-ray photographs are shot, before the dye is washed away by the blood.

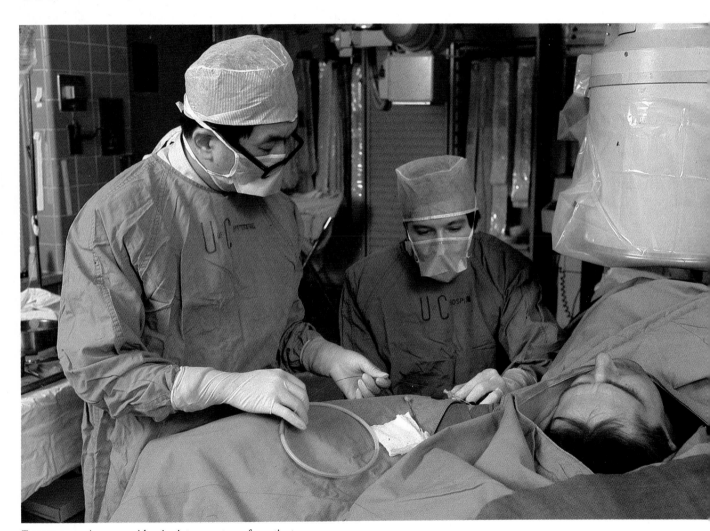

Two surgeons insert a guide wire into an artery of a patient who has liver cancer. After passing the wire through blood vessels to the diseased liver, the doctors can slip a tube, or catheter, over it. After the wire is withdrawn, the open catheter can serve as a conduit for dye. When injected, the fluid reveals the circulatory system of the liver on an X-ray film (opposite).

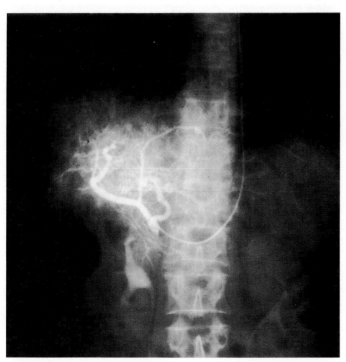

The irregular pattern of blood vessels in the patient's liver (left of the spinal column) is mapped by dye. The catheter is seen here as a thin line curving across the spinal column; the thick line at the end of the catheter is the liver's main artery. This liver is not bleeding, but the disorganized branching of its circulatory system is a sign of cancer: Blood is being diverted by the tumor.

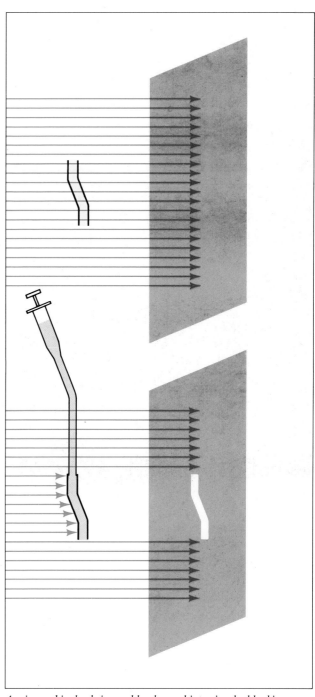

Angiographic dye brings a blood vessel into view by blocking X-rays, which are indicated by horizontal lines in these diagrams. At top, X-rays pass through an undyed vessel and uniformly blacken the film. At bottom, a blood vessel filled with dye absorbs the X-rays, leaving an unexposed white area—its silhouette— on the film for doctors to study.

Echoes to see the organs

Digestive organs are so soft that they are invisible to an X-ray machine, but there is a way to see them: Sound waves, so high pitched that they are inaudible, can be bounced off the organs and the echoes displayed on a television screen *(below)*. The technique that bares the intimate details of these organs is called ultrasonography, an adaptation of the sonar that detects submarines underwater. Its specialty is revealing gallstones, although it is also useful for finding cysts, tumors and cirrhosis of the liver.

The procedure is painless, simple and safe. A transducer, the instrument that both transmits and receives the sound waves, is passed over the organ. The sound waves travel through tissues of the organ until they bounce off a tissue of a different density, such as a gallstone, cyst or the outer border of the organ. The echoes are then converted into electrical impulses and shown on a viewing screen as murky light and dark regions, representing healthy or diseased tissue. Because air reflects ultrasound almost completely—so different is its density from that of surrounding tissue—the air-filled stomach bounces back all sound sent into its area, blocking any view of that region.

An ultrasonogram can reveal a tumor but cannot tell if it is malignant. Once the tumor is detected the physician can insert a needle into the growth and draw out tissue to be analyzed.

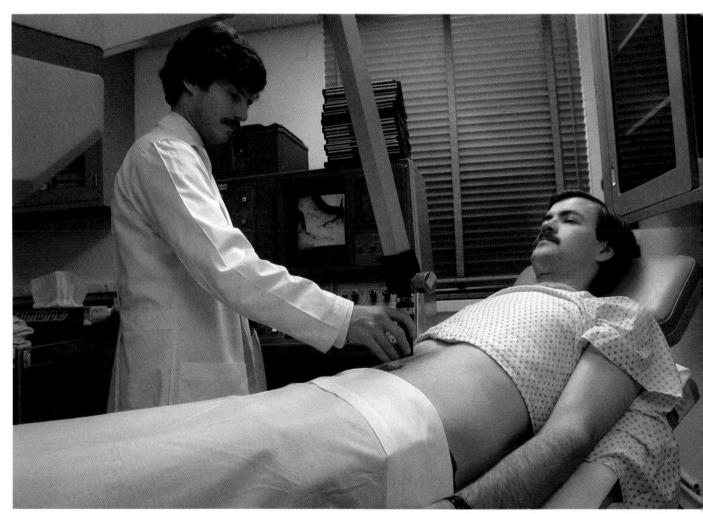

Sliding an ultrasonic transducer across the skin directly above the liver, a physician demonstrates how the device gets a sonar-like view of the organ. The sound echoes picked up by the transducer are converted electronically into an image on the screen in the background—the liver fills the screen.

Guided by the ultrasound image on the screen in the background, a physician can draw a test sample of tissue from the liver with a needle and syringe, as demonstrated at right. Each tissue sample—as many as 15 may be needed—can be extracted in about five seconds. The brown spot near the needle is antiseptic.

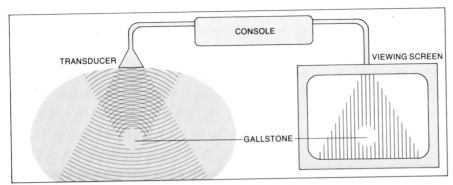

CONSOLE

TRANSDUCER

VIEWING SCREEN

GALLSTONE

Inaudible sound waves are transmitted into an organ by a transducer. When the waves encounter tissue of different density, such as a hard gallstone that is surrounded by fluid, they bounce. The echoes received by the transducer are processed in the console to make an image of the gallstone on a screen.

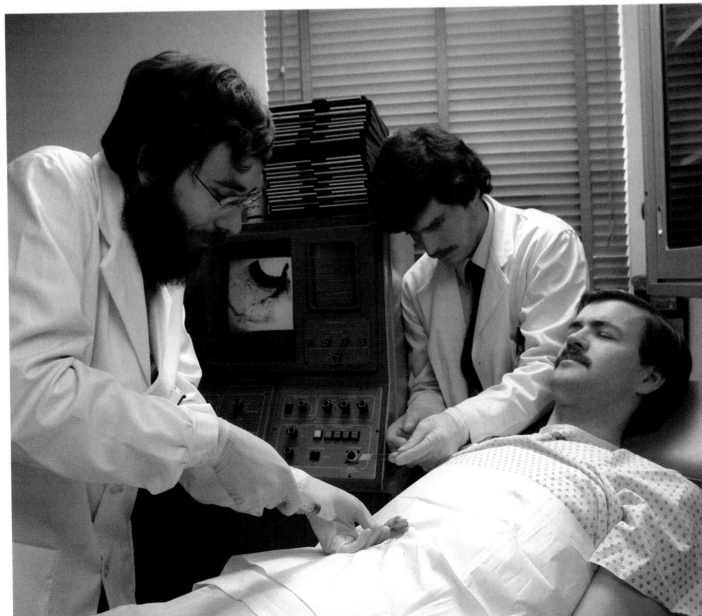

Radioactive meals to test stomach efficiency

How fast the stomach gets rid of food is one indicator of digestive health. If food passes too slowly, the stomach may have a blockage, or there may be inflammation lower down. If the stomach empties very fast—a condition called dumping, which sometimes is a side effect of surgery—the result can be persistent diarrhea.

To measure stomach efficiency accurately, physicians mark food with low-level, harmless radioactivity so that they can follow it and measure its volume as it progresses through the system. The more food that is present in each part of the tract, the more radiation emitted from that part and the darker its image on film.

To time the movement of the food, a photographic record is made every 15 minutes for up to one and a half hours. From these images a curve of the patient's emptying rate is plotted and compared with a normal emptying curve. After 45 minutes, half of the radioactive meal should have left the stomach.

If dumping is the problem, it can generally be controlled with a diet that eliminates fluids at mealtimes and is low in carbohydrates, which pass quickly through the stomach, and rich in more slowly digested proteins and fats. If the stomach retains food too long, the doctor may have to check further. When the cause is simply weak stomach contractions, drugs will help, but a mechanical obstruction usually calls for surgery.

Performing the first step of a stomach-emptying test, a medical technician injects a tiny dose of radioactive technetium into a patient's breakfast bowl of oatmeal. Eaten with the oatmeal, the radioactive tracer emits rays that will reveal to a scanning apparatus (opposite) how long the meal stays in the stomach.

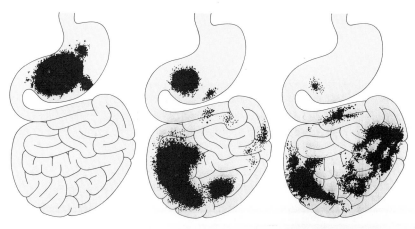

These recordings from a stomach-emptying test, retouched for clarity within an outline of the digestive tract, show a meal in normal transit. Five minutes after being eaten, the meal is entirely within the stomach (left); after 45 minutes, half the meal has passed into the intestines (center); and after one and a half hours, the stomach is virtually empty (right).

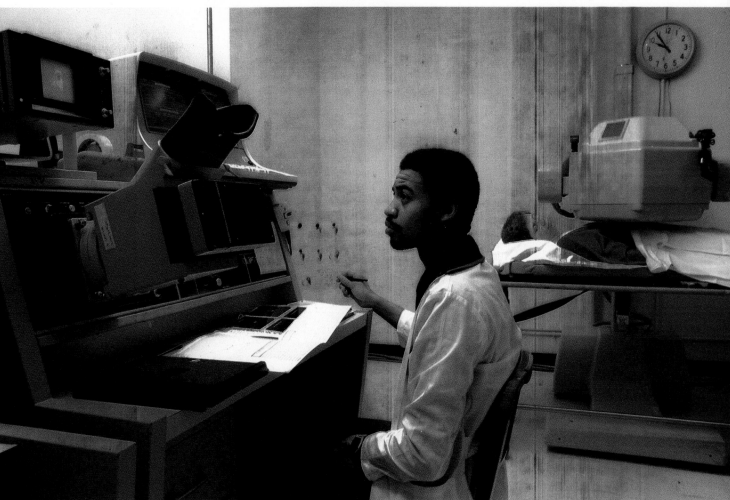

As the patient lies beneath a giant scanner, the technician monitors the progress of her radioactive breakfast, watching images on the green screen that are created by electronic detectors in the scanner. At intervals he presses a button to record the radioactivity on film, producing pictures like the ones at top.

Assays to check body chemistry

The digestive system is a kind of chemical factory, and when it goes awry it produces an abnormal batch—the wrong kinds of compounds, or too much or too little of the right kinds, or the right kinds in the wrong places. When doctors can find out which chemicals are present where and in what quantity, they often can deduce the source and severity of the trouble.

For example, heartburn, a burning sensation in the chest usually produced by stomach acid in the esophagus, can sometimes be far from trivial, involving persistent, severe pain. To diagnose and treat such cases, doctors need an objective measure—something to tell them whether acid is present, and if so, how concentrated it is. To test for this, the physician uses a probe *(drawing, opposite)* lowered on a thin tube into the esophagus via the nasal passage.

Even a patient's breath offers chemical clues to ills. A simple breath test proves whether the milk sugar, lactose, is causing chronic indigestion. First the patient is fed a test solution containing lactose. At intervals a breath sample is collected in an inflatable bag and analyzed by a gas chromatograph, which sieves out the various gases as they pass through a long tube of fine powder, indicating the volume of each gas on a graph *(opposite)*. If the patient does not digest the test solution properly, the gas chromatograph will reveal unusually large amounts of hydrogen gas.

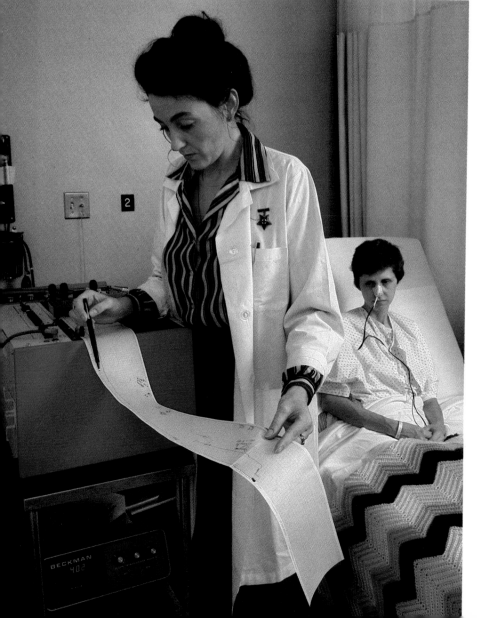

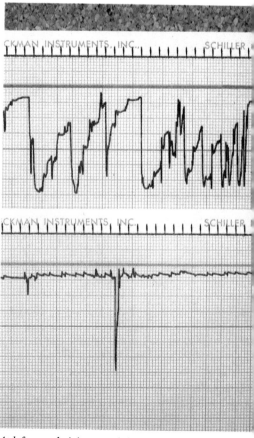

At left, a technician scrutinizes a record of esophagus acid. With excess acid (top graph, above), a tracing dips repeatedly below the neutral red line; in the lower, normal graph, it lies mainly in the alkaline zone above the line.

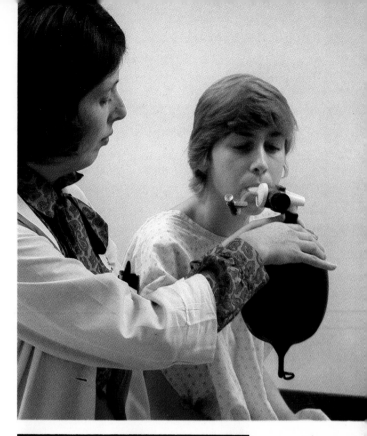

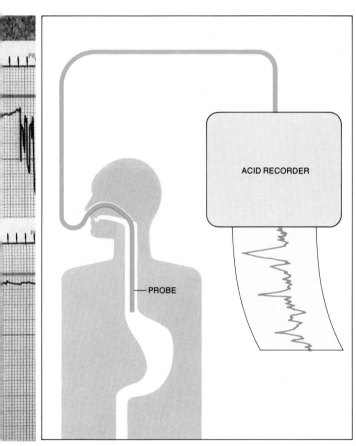

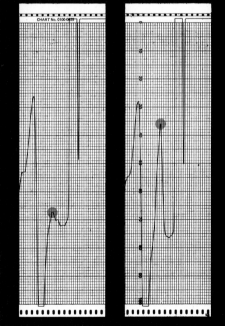

Inserted through the patient's nose, a probe gauges the acidity in the lower esophagus. It works on the same principle as a gardener's pH meter, measuring the electrical voltage that is generated when fluids in the esophagus come into contact with the materials of the probe. This voltage, a direct indication of relative acidity, is recorded on a graph.

At top, a physician draws a breath sample from a young man who may be unable to digest milk. The sample will be analyzed and graphed by a gas chromatograph. The machine produces traces such as those above, in which the peaks marked by colored dots are hydrogen—normal in the left graph but at right indicating milk intolerance. The two other peaks are oxygen and nitrogen.

Adding up the answers

Once the innovative machines have completed their work testing patients for digestive ailments, old-fashioned methods take over. Chief among these is a procedure that was instituted in most university hospitals around the turn of the century: the grand-rounds conference.

Each week at The University of Chicago Medical Center, resident physicians training there review their gastroenterology cases in a critique led by their professor, Dr. Joseph Kirsner, an internationally known authority. The progress of each patient in the department and the results of all tests—endoscopy, angiography, ultrasound and many others—are critiqued, and crucial questions are settled in the discussions. Here decisions may be made on the course of therapy to pursue, whether to order more tests, what drugs to prescribe and whether to plan surgery.

Dr. Joseph Kirsner gestures toward X-rays of the center's patients during a review of cases with resident physicians; Dr. Paul Frank (left background), Director of Diagnostic Radiology, helps interpret the photographs. This conference was spirited; Dr. Kirsner expressed plain-spoken dismay that one patient had been allowed to break his diet and consume a doughnut and soda.

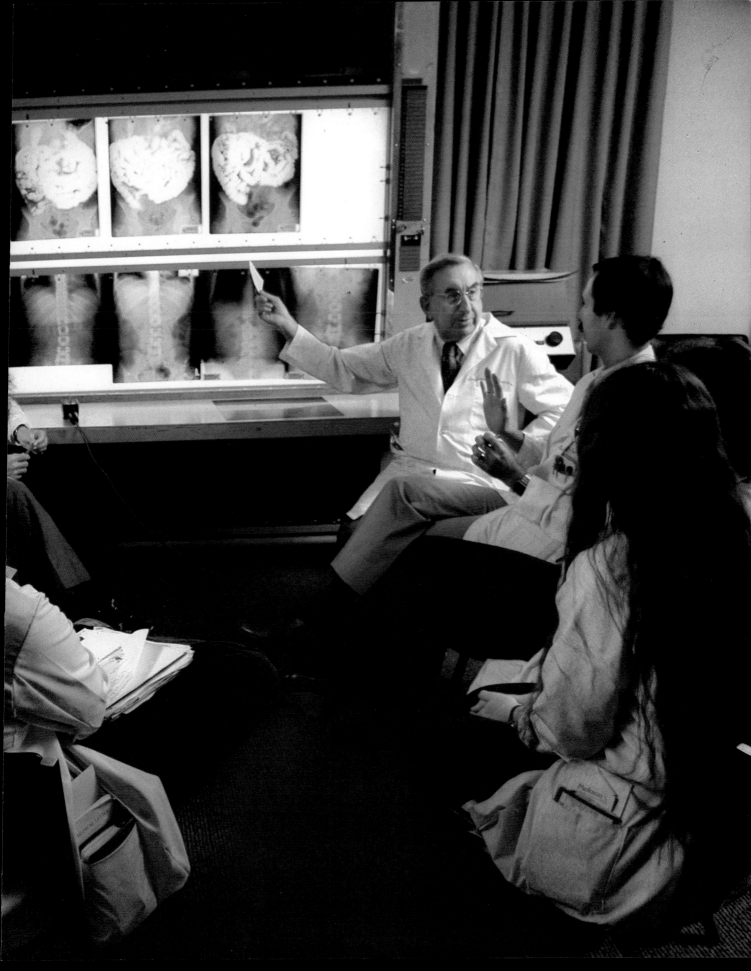

CHAPTER 5

Fighting the big-time troublemakers

Hepatitis: attack on the organ of 500 functions
New cures for the agony of gallstones
Replacing the irreplaceable kidneys
Diverticular disease: a penalty of aging
IBD: a curse on the young
A bewildering variety of treatments that help

Among the many ills that can beset the digestive system, a few major disorders bring more hardship and distress than all the others put together. Not only can each ailment be debilitating and protracted, but full recovery may be elusive. And in some cases, complications endanger life itself.

Although the small and large intestines can be affected, the potential digestive killers strike mainly at organs that are not integral parts of the tract but are associated with it: the pancreas, the kidneys and particularly the liver. Indeed, diseases of the liver are the eighth leading cause of death in the United States; one principal manifestation, viral hepatitis, strikes more than half a million people each year.

Malfunction of the liver may also generate gallstones—less dangerous than hepatitis but far more prevalent. At least 16 million Americans have them, and surgery for gallstones is the most common major operation on the digestive tract. Gallstones, in turn, influence disease in the pancreas. And kidney disorders are almost as widespread as gallstones.

The most common intestinal ailment is diverticular disease, the little pouches in the colon that develop in a third of all Americans over 45 and more than two thirds of those over 60. Generally dormant and benign, the condition often goes undetected but can sometimes flare up into alarming infection. A final type, the inflammatory bowel diseases, are—after cancer—the most serious and disruptive of illnesses of the intestines, affecting some 1.2 million Americans and costing more than $200 million a year.

Cancer, of course, is the most dangerous of the afflictions of the digestive system. It strikes the stomach rarely in the Western countries but more often in Japan—apparently diet is a factor. Cancer of the colon is tragically common in the West, attacking more than 100,000 Americans a year (about half can be cured). Cancers of the pancreas, liver and kidneys are less common but difficult to treat. Malignancies, however, are in one sense not part of the digestive system but rather unwanted additions to it; varied as they are, they share basic similarities in character and treatment and make up a category of their own, separate from malfunctions of digestion.

In advanced stages, the severe maladies of the digestive system and its allied organs cause painful inflammation of the tissues, and may bring permanent organic change to the body. All are accompanied by mysteries and obscurities that fascinate and puzzle the medical experts. But thanks to new methods of treatment and devices for diagnosis—particularly the various endoscopes *(right),* which enable doctors to look inside the digestive tract—all these ailments can generally be controlled, especially if caught early. And each year brings ways to handle them better.

That the liver can malfunction should surprise no one, for it is an organ of almost unbelievable complexity and versatility. It serves at least 500 roles in the operation of the human body. Besides manufacturing bile—that critical elixir for digesting fats—it produces proteins that help blood to clot and others that enable the immune system to protect the body against invading agents. It also makes some essential cholesterol, converts carbohydrates into fuel for muscles and regu-

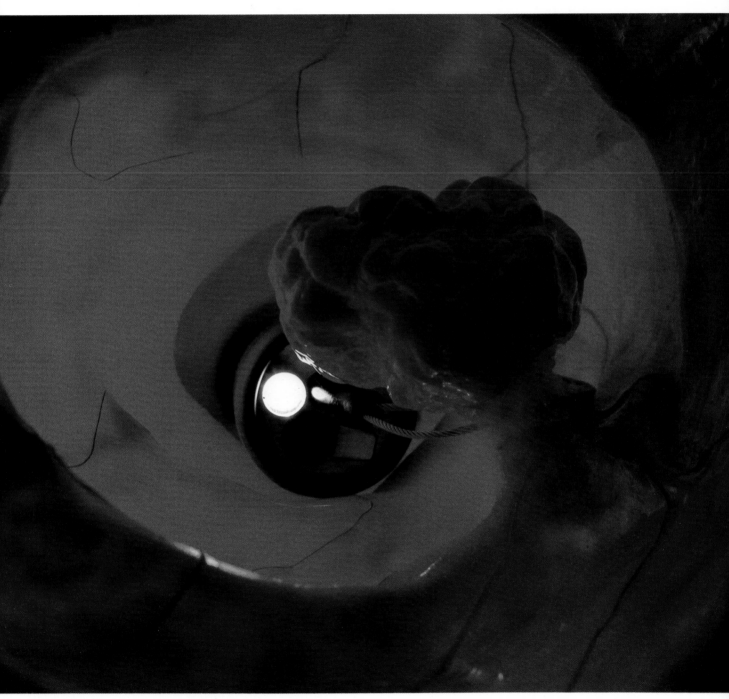

*How a colonoscope snares a potentially cancerous growth
in the intestine—as far as five feet up the tract—is demonstrated
above in a mock-up. Like other medical scope devices (pages
92-93), it has two light pipes—one for illumination, one for
viewing—and controls for attachments.*

lates the hormones that control growth and sex. It is the only part of the body, aside from the skin, that can almost entirely regenerate itself if injured; it will grow back to normal size after as much as 70 per cent has been removed.

Perhaps most important, the liver is a purifier, cleansing the system of various toxic chemicals and bodily wastes. Herein lies the liver's major vulnerability, for it can become overloaded and unable to cope with certain harmful materials. Then the purifier itself succumbs to impurities and becomes diseased, harmed beyond its ability for self-repair. Moreover, the liver can also be damaged, sometimes permanently, by viral infections, by imbalances in body operation and even by the effects of illness elsewhere in the body—for example, by the inflammatory bowel disease that is itself a serious ailment.

Hepatitis: attack on the organ of 500 functions

The most common harm to the liver comes from hepatitis— the name merely means inflammation of the liver. There are two main types, viral and toxic. Viral hepatitis is caused by virus attack on the liver; toxic hepatitis is the cumulative deterioration of liver cells by chemicals—alcohol, drugs and industrial compounds. If either kind of hepatitis is not treat-

Swathed in sterile garb, a technician begins the dangerous task of harvesting virus from blood for a vaccine against the sometimes fatal liver disease hepatitis B. The 65-week production process, which starts with pouches of straw-colored plasma (above) taken from carriers of the disease, inactivates the live virus but preserves viral particles that provide immunity when injected.

ed, it may cause the liver to shrink and harden into the condition called cirrhosis.

Several kinds of viruses strike the liver. In hepatitis A, traditionally known as infectious hepatitis, the virus is passed from person to person, usually through fecal contamination of food or water, spreading in crowded places with inadequate sanitation facilities, such as army barracks or urban slums; a contaminated meal in a greasy-spoon restaurant, or a dish of polluted shellfish can also bring it on. Hepatitis B is sometimes called serum hepatitis, because doctors once thought it was spread only by injection of infected blood or blood serum, or by unsterilized hypodermic needles such as those used by drug addicts. But the virus can also be transmitted in food or water. Doctors have now found other strains of viruses that harm the liver and, like the B virus, spread mainly through faulty transfusion procedures; somewhat confusingly, this group of agents is called non-A non-B.

Viral hepatitis is more prevalent than liver disease caused by consumption of toxic substances. Among the compounds that, if ingested or inhaled to excess, can bring harm to the liver are not only alcohol but also medicines such as some used to treat blood pressure or even headaches. Many volatile industrial compounds, which contaminate the air in workplaces, have also been implicated. The once-common solvent carbon tetrachloride is an example. It was found to make liver damage such a serious occupational hazard of dry cleaners that its use for cleaning was banned.

Despite the abundance of terms and causes for hepatitis, its many forms produce much the same symptoms and are treated more or less the same way. The differences are in details.

Hepatitis is likely to manifest itself first through loss of appetite, nausea, low-grade fever and occasionally pain in the upper abdomen—produced by swelling of the liver. It also affects the sense of taste—cigarette users suddenly find smoking unpleasant. Jaundice, the classic warning of liver trouble, may follow, but it is not a certain sign. It appears in only about one case in three. Jaundice is not a disease in itself but merely a symptom. Recognizable by a yellowing of the eyeballs and skin, jaundice results from a backup of bile pigment into the bloodstream; the pigment, known as bilirubin, cannot leave the body by way of its normal route through the liver and intestines, and comes out instead via the kidneys. Thus the urine may turn dark and the feces can be light.

Sometimes hepatitis in its early stages will mimic flu and cause the victim simply to take to bed, but the weird look of jaundice—often first noticed by a friend or relative—will bring a quick call to the doctor.

Jaundice's telltale yellow blush usually permits swift diagnosis of hepatitis. In serious cases the doctor may then go on to distinguish among the forms. This information will indicate how long the patient will be sick and whether the liver is likely to suffer permanent damage. The doctor can often make an educated guess on the basis of the patient's habits and life style—housing conditions, use of alcohol—but blood samples provide more certain identification. Another diagnostic aid is a liver biopsy, in which a hollow needle is inserted through the abdominal wall into the liver to obtain a tiny bit of tissue for analysis.

The infections caused by the various viruses can usually be distinguished by degrees of severity and by characteristic periods of incubation—the time between exposure to the virus and onset of illness. For Type A, the incubation is two to six weeks, while for Types B and non-A non-B it may be anywhere from six weeks to six months. Hepatitis A can be so mild as to be unnoticeable, and can pass as easily as a cold or a case of indigestion. But it can also, like Type B, put its victim on his back for weeks; he may be barely able to hold down food or even to walk across the room without collapsing. Types B and non-A non-B are more dangerous. They can be transmitted by carriers who are unaware of their own infections; they cause more severe, even fatal, illness; and they occasionally cause permanent damage to the liver.

Depending on identification of the virus strain, the physician may want to inoculate close friends and family members in hopes of preventing its spread. To be sure, spread of hepatitis A is preventable through simple sanitary measures: Hands should always be washed after bowel movements and before meals. But so contagious is this type once it is present that physicians frequently advise injections of gamma globulin, a protein that is formed in the blood and that has proved

especially effective in preventing hepatitis A—it supplies antibodies that kill the virus. Gamma globulin is often administered to hospital and other institutional workers who must be exposed to hepatitis, and can even be given to travelers who plan to visit areas where sanitation is primitive.

Although gamma globulin is sometimes given to ward off hepatitis B, the results are somewhat less certain. Greater assurance comes in the form of a vaccine made from blood from hepatitis B carriers. When injected into a healthy subject, it stimulates the body to make antibodies capable of preventing the illness; tests indicate that it is almost completely effective. The non-A non-B type is much more difficult to protect against because it apparently is caused by not one but several distinct viruses.

As it happens, there is little that doctors can do—or need to do—to treat any type of viral hepatitis once it is contracted. Antibiotics have no effect. The patient is merely instructed to take it easy. And he must scrupulously avoid alcohol, which is particularly injurious to an ailing liver. Patients are advised to stay home; their eating utensils and bed linens should be kept separate from everyone else's. In most cases the disease causes minor discomfort—many patients mistakenly dismiss it as influenza. But it can be severely debilitating. And the infection takes about three months to clear up.

Many of the toxic forms of hepatitis follow a course similar to that of the viral disease. But the toxic hepatitis caused by alcoholism, the most common form, generally progresses in three stages. In the first stage, alcohol disrupts the liver's processing mechanisms, causing fat cells to proliferate; the liver grows larger. The early symptoms may be a mild pain on the upper right side of the front of the torso and, occasionally, a tinge of jaundice. The second stage is alcoholic hepatitis, bringing symptoms similar to those of viral hepatitis—nausea, loss of appetite and so on. At this point a proper diet and a rigorous taboo on liquor will generally provide a cure; unfortunately the confirmed drinker all too often refuses to stop. So the third stage ensues.

Under continued and protracted onslaughts of alcohol, the liver begins to deteriorate rapidly, becoming hard, fibrous and somewhat shrunken; scar tissue obstructs the flow of blood, and liver functions are drastically impaired. The palms of the hands turn red and the legs and fingers swell up. Blocking of the liver's blood vessels may cause fluids to escape into the abdominal cavity and the sufferer, though malnourished, may develop a potbelly. As with viral hepatitis, there is no specific medical remedy.

New cures for the agony of gallstones

A disorder quite different from hepatitis comes about when the liver somehow fails to maintain the proper balance of chemicals in its secretion of bile. The liver itself may not be harmed, but other parts of the digestive system often are.

Bile is a biological cocktail of pigment, cholesterol, and various salts and acids dissolved in water. Most of the bile is shunted into the small, pear-shaped gall bladder for temporary storage. There it becomes highly concentrated, and this brings the most common problem. Normally, the cholesterol and pigment remain in solution. But if the bile becomes supersaturated with chemicals, some precipitate out to form small solid lumps—gallstones. Two bile components are involved: cholesterol, which forms round, light-colored stones if bile acids decrease, and the bile pigment bilirubin, which forms dark flaky stones for reasons unknown. The two kinds occasionally form in combination, but most stones

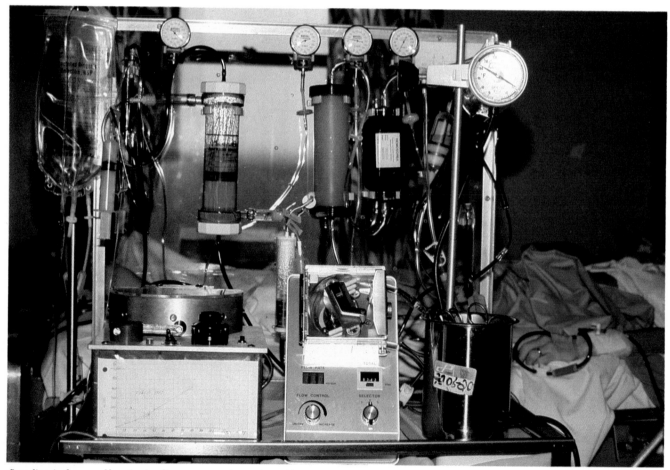

Standing in for a malfunctioning liver, an experimental machine filters toxins from blood drawn from a patient's arm (lower right). First, plasma is separated from the blood by the cylinder at left; toxins in the plasma are then removed by the cylinder and black container at right. The purified plasma is reunited with the whole blood and returned to the patient.

are mainly one type or the other. They can appear in the liver or gall bladder or in any of the ducts that connect these organs to the intestine. Stones may be no bigger than a pinpoint, but they commonly grow to about a quarter of an inch in diameter; some have been measured at three inches across.

The stones appear because the bile is constantly changing: It is very concentrated at some times, dilute at others; when it is overloaded it dumps some of its dissolved constituents in solid form. The possible causes of overload are many. Disease is one: Blacks afflicted with the blood ailment sickle-cell anemia are prone to gallstones. Drugs are another cause: Women who take contraceptive pills—which alter the balance of hormones—double their risk of gallstone attack.

Diet is also involved: Gallstones are notably common among people who, at the urging of heart experts, follow a low-cholesterol diet and eat unsaturated vegetable fats instead of saturated animal fats. Some drugs that lower body cholesterol are also associated with increased incidence of gallstones—apparently the cholesterol removed from the blood is excreted into the bile, increasing its load. And in Japanese cities, where people have turned to a Western style diet high in fat, the incidence of cholesterol gallstones doubled over 47 years while the incidence of pigment gallstones went down; in rural villages, where diet changed little, the gallstone types changed little.

But the most striking example of the effects of living habits is found among the Indians of the American Southwest. In the 19th Century the Pima, who live in Arizona southeast of Phoenix, rarely suffered from gallstones. By the 1980s, almost three fourths of Pima women over 30 were afflicted—apparently one consequence of a shift from tortillas, chili and beans to the fatty foods of modern America.

Age, sex and heredity also seem to determine who gets gallstones. Stones run in families, and they develop in four times as many women as men.

The stones can occur at any age. Children get them, and a significant portion of victims are in their twenties. Many an octogenarian has been hit. If gallstones remain in the gall bladder they may never cause trouble. In fact, out of the estimated 16 to 20 million people in the United States who are believed to harbor stones, perhaps half have "silent" ones that will never produce symptoms. But in some 11 million Americans the stones are far from silent. Escaping the gall bladder, a stone or two will lodge either in the cystic duct or, farther along toward the intestines, in the common duct.

If the stone is not so big as to obstruct total bile flow, the result will be mild pain. But if the stone is large enough to cause total stoppage, the victim is likely to feel sudden and protracted pain in the upper right quadrant of the abdomen. It can be agonizing. "I've felt pain before," said a New York executive who was struck during the night while on vacation in Bermuda, "but nothing like this." Unable to excrete its bile, the gall bladder becomes distended and inflamed. So may the bile ducts. The pain may be accompanied by nausea and vomiting, and jaundice may develop. The pain is unwavering and lasts perhaps an hour or so, then subsides gradually to become a mild ache. It may or may not return, but the victim will not forget what happened.

Some physicians may be able to diagnose the ailment merely by examining the patient. The executive stricken in Bermuda recalls that the local doctor, after listening to his story and feeling his abdomen, announced impeccably and with total assurance, "You have gallstones." He happened to be right. But the chances of error are considerable. The same symptoms might mean a perforated ulcer, heart attack, acute pancreatitis (an inflammation of the pancreas) or even appendicitis. The usual recourse is to X-ray. But although gallstones of the less common, pigment, variety tend to show up on ordinary X-rays, the others do not. So the diagnostician may order a second set of pictures in which a dye opaque to X-rays is used to outline the gall bladder and bile ducts.

X-rays are increasingly giving way to ultrasonography, in which high-pitched, inaudible sound waves search out the stones the way sonar scans for submarines, displaying images of the quarry on a video screen. Whatever the method, the patient may be startled by the size and number of stones.

Then he faces the question of what to do about the rock pile. Often an obstructing stone will become dislodged of its own accord and be carried into the intestines and thus out of the body; the symptoms will vanish. One 80-year-old woman was hospitalized for removal of her gall bladder after X-rays revealed more than 50 stones; by the time the operation was performed, all but one had left her body. But doctors know that once a single gallstone has made its appearance, others are likely to follow. In a study of 781 Swedish patients whose stones had passed and who had then enjoyed a full year free of complaints, researchers found that more than a third suffered renewed acute attacks within the next decade.

Generally doctors begin treatment by administering painkillers while they attempt to judge the seriousness of the disease. If it seems unlikely to disappear on its own, they will probably consider surgery. Removal of the gall bladder is a safe procedure and, except in the uncommon cases in which stones appear in the liver, the surgery eliminates the threat of gallstone disease permanently. The operation is performed some 500,000 times each year in the United States.

Until recently, surgery was the only recourse—beyond wishful thinking. Now there is a medicine that will dissolve some gallstones. In the 1960s researchers at the Mayo Clinic in Rochester, Minnesota, discovered that one of the two bile acids synthesized by the liver, chenic acid, or CDCA, could, if taken in sufficient quantity by a gallstone sufferer, alter the composition of the bile and restore the normal chemical balance. Stones of the cholesterol type, most common in Western patients, disappear back into solution. However, CDCA has no effect on pigment stones.

All through the 1970s CDCA was tested extensively; the drug itself was produced from bile obtained from the livestock industry. Finally in the spring of 1981 the results of a nationwide study were announced. Relatively large doses of the drug, reported the inquiry's director, Dr. Leslie Schoenfield of the Cedars-Sinai Medical Center in Los Angeles, completely dissolved the stones of 14 per cent of patients in the study and partially dissolved those of another 27 per cent. Side effects were, happily, few. The drug seemed to work best with patients whose stones were sufficiently developed to cause intermittent pain, but not so advanced or uncomfortable as to require immediate surgery. For people in this limited category, the chances seem bright that they may be able to take pills to control their problem.

Gallstones frequently cause pancreatitis: they can block the flow of the enzymes from the pancreas so that these active chemicals back up and begin to digest the tissues that produce them. The other principal cause of pancreatitis is alcoholism. But the pancreas generates such a copious flow of digestive juices that serious trouble is likely to arise only after 90 per cent of its cells are put out of action.

Replacing the irreplaceable kidneys

A greater hazard than pancreatitis is failure of the kidneys, the organs that must regulate the body's water supply and ultimately excrete important digestive wastes. "Bones can break, muscles can atrophy, glands can loaf, even the brain can go to sleep without immediate danger to survival. But should the kidneys fail," wrote Dr. Homer W. Smith of New York University, "neither bone, muscle, gland nor brain could carry on."

An estimated 13 million Americans suffer from kidney diseases, and 50,000 die each year. The most serious dangers are inflammation, generally the result of infection; leaks in membranes that allow blood to escape into the urine; and injury or poisoning that destroys kidney cells. Most of these afflictions, if caught early enough, can be controlled with antibiotics to eradicate infections, and hormones to help repair damaged tissues. But harm so extensive that the kidneys shut down their operation was—until World War II—inevitably fatal. Today the kidney-dialysis machine can take over. The most effective kidney machines mimic the action of the natural organ, pumping the patient's blood through plastic tubes to filter out wastes and excess water, then sending the cleansed blood back into the body.

The first successful machines were built during the German occupation of the Netherlands, by Dr. Willem J. Kolff, who used cellophane sausage casings from Germany, enamel tubs secretly made by a local factory, rotating couplings copied from a Ford water pump and wooden drums crafted by a cartwright. ("Since seasoned wood was not available," recalled Dr. Kolff later, "we had some problems with shrinking.") He treated 17 patients in two and a half years, before his invention saved the life of one—a Nazi collaborator whose death "many of our fellow citizens would have accepted," Dr. Kolff commented, "without regret."

Similar devices had been under development in many countries. By 1950, when Dr. Kolff left the Netherlands for a position at the Cleveland Clinic, he found there a stainless-steel version; it "cost $6,000, but it worked as well as the $200.00 wooden machines made in The Netherlands."

By 1979 artificial kidneys were keeping alive more than 120,000 people around the world who otherwise would have died. That year the United States government spent more than a billion dollars to provide these treatments or kidney transplant surgery *(Chapter 6)* for approximately 50,000 Americans—around $20,000 per patient.

A newer method, which uses the patient's own abdominal lining as a filter, is simpler and more economical. Called continuous ambulatory peritoneal dialysis, it does not require a machine to pump blood through a filter. Instead, a sterile solution is emptied from a plastic bag through a permanently implanted tube into the patient's abdominal cavity. Wastes and excess water pass of their own accord from the blood-stream through the patient's lining membrane into the solution, and the waste-filled solution drains back out the tube and into the bag for disposal.

Diverticular disease: a penalty of aging

Although very dangerous afflictions, such as kidney failure or hepatitis, can strike organs that are accessories to the digestive tract, the tract itself is remarkably resistant to lethal attack. When it is stricken, the outcome is seldom death—but often misery. In addition to ulcers *(Chapter 4)*, two great families of disorders trouble the small and large intestines—the diverticular and inflammatory bowel diseases.

Diverticular disease, according to one book on the subject, is "the price many people pay for surviving into middle age and older." The ailment is extremely rare among those under 30. But starting at about 30 the large intestines of many people develop weak spots, especially in the last main section, the sigmoid colon. As food residue or gas brings pressure on the colon, these weak spots can give way and create diverticula—tiny pouches like miniature hernias, bulging

out through the colon wall. The condition, called diverticulosis, causes little trouble; most people never know they have it. When it does manifest itself it may produce cramps, gas, and diarrhea alternating with constipation. More women, for reasons not known, acquire the condition than men.

Diverticulosis would be hardly worth mentioning if it always remained harmless. Those pouches, however, can become repositories for bits of fecal material; they grow infected and inflamed. The benign ailment will then have escalated into diverticulitis—inflammation of the diverticula—and it is a serious disorder. The inflammation not only hurts but frequently causes the pouches to perforate, so that infected material spreads outside the colon into the abdominal cavity; this is peritonitis, a complicating infection of the abdominal lining that, before the days of antibiotics, was often fatal.

Diverticulitis will probably thicken the wall of the colon, and may even close the tract altogether. The sufferer will be nauseated and feverish, may vomit, and perhaps will see some blood in the stools. The pain, felt in the lower left quadrant of the abdomen, is sharp. Doctors sometimes call diverticulitis "left-side appendicitis," as its symptoms are similar to that ailment in nearly all respects save location.

There are three possible causes of diverticulitis and its precursor, diverticulosis. One is simple muscle shrinkage in the wall of the aging colon, which may weaken the whole structure. Another is the use of potent laxatives. Consumed in hopes of speeding elimination, they too often have the opposite effect: Bringing artificial pressure on the colon, they may weaken its walls and make them less resilient. This loss of resilience leads in turn to more constipation and further weakening. A third contributor, many doctors believe, is the low-residue diet popular in most Western countries; they feel that foods from which most of the indigestible fiber has been removed result in hard stools that tax the colon.

Diet is so often blamed for the diverticular diseases because the marked increases in these diseases during the 20th Century coincide with dramatic changes in the kinds of foods people eat—fiber was removed from flour by new milling methods and the consumption of fiber-free meats and sugar increased. Diverticulitis was almost unknown in the 19th

An artificial kidney that can go anywhere

When Bud Jackson's kidneys gave out in the early 1970s, the Murray, Utah, draftsman felt the world closing in on him. He could never venture far, he thought, from the artificial-kidney machine that three days a week, for five confining hours at a time, circulated his blood through cleansing hemodialysis membranes, filtering out toxins his own failed kidneys could not remove. An avid angler—he had spent his vacations fishing on Idaho's Salmon River—Jackson confessed he "never expected to be able to do anything like that again." Yet in 1980, he was white-water rafting down the Salmon with his wife, a group of medical technicians and 10 other people with kidneys as useless as his own.

The trip was made possible by an experimental, battery-powered device called WAK, for Wearable Artificial Kidney. Developed at the University of Utah, the WAK is an eight-pound blood-filtering mechanism that can be strapped to the chest. Unlike the peritoneal dialysis device shown on pages 146-147, it does not require a permanent opening in the body; thus it can be used under less sanitary conditions. The WAK can be carried and set up virtually anywhere, without the permanent electrical and plumbing connections required by standard artificial kidneys—and the trend is toward even greater convenience. Said the inventor of the first successful kidney machine, Dr. Willem J. Kolff: "The time will come that all artificial kidneys will be totally wearable and so small they can hardly be seen."

A raftload of kidney patients and their companions shoots a rapid. Temporarily liberated by portable devices from the big machines that ordinarily keep them alive, the patients spent a week on a wilderness excursion sponsored by the University of Utah.

Connected by tubes in his left leg to the Wearable Artificial Kidney that hangs in the stand at his side, Bud Jackson pats a horse that wandered into the camp. Fluid in the bag at the base of the stand rinses out wastes from the WAK's filtering materials. The WAK may be disconnected from the bag for 15 minutes at a time, allowing patients to move about while their blood is cleaned.

Bud Jackson displays a whopper he hooked while attached to the WAK at riverside. Other campers took short walks along the river with their WAKs strapped to their chests.

Century; it was not even mentioned in medical textbooks published in the early decades of this century.

By the 1970s, estimated Neil S. Painter of Manor House Hospital in London and Dr. Denis P. Burkitt of Britain's Medical Research Council, one third to one half of all people over age 40 in the United States, Great Britain, France and Australia had developed diverticulosis, and at age 80 the incidence was two thirds. It had become, these researchers noted, "the commonest affliction of the colon in Western countries within 70 years, the traditional life span of man." By contrast, such ailments seemed rare among Africans and Asians, who live on a traditional diet of whole grains and vegetables high in natural fiber. One study over five years in Singapore found only one case per 150,000 Asian residents in the city—but one case per 5,000 Europeans there.

This evidence implicating low-fiber diet as a cause of diverticular disease is circumstantial. But it is reinforced by experiments on rats and rabbits, which developed contracted and pouched colons when fed little fiber.

Research on a great many people over a great many years will be necessary before the link to diet is firmly established. But in the meantime, most authorities follow a common-sense course: Because fiber-containing foods may help prevent diverticular diseases, they recommend menus containing such fibrous foods as fresh fruits, whole-grain cereals and green leafy vegetables.

When diverticular disorders occur, only about 12 per cent of the patients develop diverticulitis, but the ailment must be treated aggressively. A blood count and a "lower GI" series—X-rays of the lower part of the gastrointestinal tract made with the help of a barium-containing enema—aid diagnosis. The doctor may perform a direct examination of the area with a colonoscope, a type of endoscope (pages 94-95) that gives a view of the mucosal wall. He may also use the colonoscope to take a biopsy, as a double check for cancer.

Diverticulitis, despite its threatening nature, can usually be treated successfully at home. Patients will be ordered to bed and, for the time being, will be put on a low-residue, bland diet or even a liquid diet—this is not the moment to challenge the colon with roughage. Antibiotics should clear up the inflammation and infection. Gradually, as symptoms abate, the doctor will reintroduce fiber-containing foods.

In truly severe cases, diverticulitis patients may need to be hospitalized and given drugs to slow down intestinal activity. If an infected pouch bursts and threatens peritonitis, if the colon has become obstructed, or if a blood vessel has been ruptured, surgery may be called for. Fewer than 10 per cent of diverticulitis patients require surgery, which is avoided partly because it requires two and sometimes three separate operations, over several months.

IBD: a curse on the young

While diverticular disease is commonly a disorder of older people, and is relatively easily controlled or reversed, an entirely different picture is presented by the final large category of intestinal ailments—inflammatory bowel disease, or IBD. Its primary victims are young, most often between their teens and 30s but sometimes even younger; in the United States 200,000 children under the age of 16 suffer from IBD. Its successful management is exceedingly difficult, and its cure is elusive. Moreover, it seems to be on the increase throughout the world. One of its forms, Crohn's disease, is appearing more and more in the United States, Sweden, Norway, the Netherlands and Britain—according to one estimate, incidence has increased tenfold over 20 years.

Many people suffer secretly from IBD, for it is an unfashionable, socially embarrassing ailment whose principal sign is likely to be something hardly mentionable: bloody diarrhea. The general populace becomes aware of it only rarely, as when President Dwight D. Eisenhower came down with a variety of it in 1956 and again in 1969. Tragically, its embarrassing nature keeps many sufferers from visiting a doctor until the symptoms are painfully and irreversibly advanced. If treatment can start early enough, the outcome is likely to be far less agonizing.

Its possible causes are unclear and its many possible therapies are undergoing constant revision and reassessment. "There is no firm handle by which to grasp even the fundamental nature" of the disorder, confessed one authority, Dr. David B. Sachar of New York's Mount Sinai School of

Medicine. Reflecting the difficulty is the name itself, an umbrella label that doctors have settled on as being vague enough to take in all the perplexing, shifting characteristics they keep trying to understand.

IBD is thought to consist of two major disorders, ulcerative colitis and Crohn's disease. Of the two, ulcerative colitis has been identified for a much longer time—ever since the middle of the last century. Almost invariably it starts in the rectum and works its way back. Sores or ulcers may begin to climb the sigmoid colon, attacking the mucosa—which turns fiery red and bleeds readily. The disease may stop there, or it may go on to envelop most or all of the large intestine. At any time the patient may go into remission and the ulcers may heal; but scars will remain, giving the mucosa a pitted quality that makes it susceptible to renewed injury.

As the ulcers start forming, the symptoms begin: intermittent bouts of bloody diarrhea and moderate cramping. Often there is fever. The symptoms may come on gradually, or all at once. The nine-year-old son of novelist Inge Trachtenberg was the picture of health one day; the following day he came down with agonizing diarrhea, and cramps so severe he was initially thought to have typhoid fever. At the height of the disease's fury a patient may have 10 to 15 bowel movements a day and will sometimes seem never to leave the bathroom. Patients may suffer from anemia, lose weight and become alarmingly weak. One young man became accustomed to seeing his temperature hover between 102° and 104° F. for days on end. Sometimes, unaccountably, there will be arthritis and skin and eye lesions.

For nearly 50 years, these ulcers of the lower bowel were all that anyone knew about. Then in the early years of this century, in both England and the United States, another kind of inflammation began appearing that was similar—yet different. No one defined it beyond thinking that it was perhaps an intestinal version of tuberculosis—there was a strong similarity in appearance to that disease. Finally in 1932 Dr. Burrill Crohn of New York's Mount Sinai Hospital and his colleagues reported that, at the lower end of the small intestine, in the section known as the ileum, they had identified patchy ulcerations like those that cropped up in the colon.

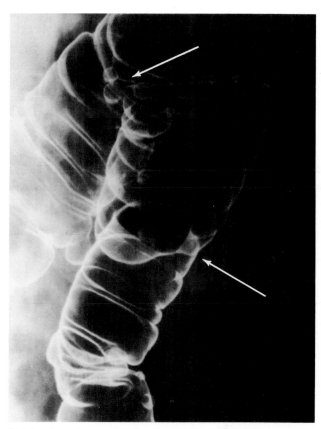

Bulging from the segmented wall of a colon, the light protuberances (arrows) seen in this X-ray are diverticula. The pouches, common in elderly people, form when food or gas presses outward against weak intestinal walls. Though normally harmless, the pockets can become inflamed, producing the pain of diverticulitis, or they may burst and infect the abdominal cavity.

Gunvor Rosén of Sweden sits surrounded by one day's food— enough for eight normal people: 15 eggs, 6½ pounds of potatoes, 4½ pounds of pork and liver, a package of bacon, 4 steaks, 12 beef slices, 2 quarts of ice cream, a pound of butter, several loaves of bread and 20 quarts of tea and beer. Because her digestive system is largely lost to Crohn's disease, she must eat huge amounts or get her nourishment intravenously (pages 122-129).

But these went deeper. Where those of ulcerative colitis were always superficial, these would eat down into the intestinal lining, sometimes penetrating it and creating an opening into the abdominal cavity. Occasionally the holes might bore through to a neighboring loop of the small bowel or to other organs—or even to the outside of the body. These tunnels, known as fistulas, did not occur in ulcerative colitis. They were serious. Dr. Crohn and his associates called the disease terminal ileitis—inflammation of the end of the ileum.

At first it was believed that this new ailment was limited to the ileum. In the 1950s, however, physicians discovered deep, ileitis-type ulcers elsewhere—in the large intestine. Later on, ulcers like those identified by Dr. Crohn were found as far up as the esophagus and even the mouth. Doctors have now lumped the whole deep-ulcer syndrome, wherever it occurs, under the name Crohn's disease. They still cannot understand how the ailment, unknown a century ago, has seemed to develop so fast. "One cannot escape the conviction," said Dr. Henry D. Janowitz, of the Mount Sinai School of Medicine, "that this really is a new disease, which emerged in the early part of this century." Otherwise, he argued, the great European pathologists of the 19th Century would surely have discovered the telltale ulcers in autopsies.

The symptoms of Crohn's disease are similar to those of ulcerative colitis; there is some bloody diarrhea (though not as much when the disorder affects only the ileum), as well as severe cramping and weight loss; there may also be skin and eye abnormalities and pain in the joints. Patients may be affected mildly or severely, and remissions can come along in random fashion. However, both diseases, once contracted, generally become chronic. "Think of it as a chronic *vulnerability*," said the national Vice President of the American Digestive Disease Society, Jane Gross, who herself suffers from Crohn's disease. "You may go for years in total remission. But your gut is a permanent weak point."

A frustrating hunt for causes

As many as a third of all cases of IBD have occurred in families where at least one other member had also contracted it. Some families are extraordinarily unlucky: Three years after Inge Trachtenberg's son came down with ulcerative colitis, her daughter, who was just three years younger, developed it too. But no genetic "markers" or identifiable transmitting agents have been found, and an overwhelming majority of IBD sufferers go on to have children who are untouched by the disorder. Remarked Dr. David Sachar of the Mount Sinai School of Medicine, "We know it is familial—but that does not mean it is genetic." The common factor could be environmental—outside influences may work on some families because of where or how they live.

In direct contrast to the genetic theory is the idea that IBD may be caused by a virus or some other infectious agent. In experiments at St. George's and Brompton Hospitals in London, tissue taken from ulcerative-colitis patients was injected into rabbits; they thereupon developed a malady similar to IBD. And when tissue from the diseased rabbits was injected into yet another group of rabbits, these too acquired the ailment. The process took many months, suggesting that the culprit was a so-called slow virus, one that produces symptoms of disease after a long period of incubation. This idea is supported by some evidence from human cases. Two sisters who had not seen each other in six years, and who lived on different continents, simultaneously contracted Crohn's disease. So close was the timing that only a virus with a precise incubation period seemed to explain it.

Another notion is that IBD, like rheumatoid arthritis and a number of other systemic disorders, may result from a breakdown in the victim's immunological defenses: Antibodies that normally fight a disease end up attacking the host body instead. Studies in Europe, the United States and Israel showed that abnormal antibodies do exist in the blood serum of ulcerative-colitis patients, and that IBD sufferers tend to have disease-fighting white cells destructive to the colon. Supporting the theory is the fact that when certain immunosuppressive drugs—medicine that tempers the body's ability to fight disease—are given to IBD patients, they tend to improve noticeably. But many doctors believe the answer lies in a combination of factors. "If further work places these clues on a more solid basis," said Dr. Michael J. Goodman of Bury General Hospital in England and Dr. Marshall Spar-

berg of Northwestern University, "then one may anticipate the finding that ulcerative colitis results from a genetically determined immunological reaction to a viral agent."

Hovering over all these considerations is the question of a psychological cause or influence. Most doctors agree that a high proportion of IBD patients exhibit certain consistent personality traits that would seem to make them especially vulnerable to emotional stress. Dr. George Engel of the University of Rochester listed "obsessive-compulsive character traits, including neatness, orderliness, punctuality, conscientiousness, indecision, obstinacy and conformity." Sufferers have an "almost uncanny perception of hostile or rejecting attitudes in others" and "tend to brood and withdraw." With rare exceptions, Dr. Howard Spiro wrote, IBD victims seem particularly unable to express strong emotions; according to Dr. Engel, most are the offspring of mothers who are domineering and on whom they feel totally dependent.

In the forefront of enthusiasts of the psychosomatic theory is a cohort of psychiatrists who are, for the most part, disciples of Dr. Melitta Sperling. A spirited, contentious psychiatrist, Dr. Sperling over a period of many years beginning in the 1940s treated 33 children who had ulcerative colitis; she claimed to have produced permanent remission in 30 of them. "Ulcerative colitis for which no known external or internal etiological cause can be established medically," she wrote, "indicates the presence of a psychiatric disorder." Only by facing up to the conflicts that produce such a disorder, she and her followers have said, can a patient gain ascendancy over IBD. Then the symptoms will drop away.

To many doctors, the trouble with this approach is that recent experience seems to indicate otherwise. Mary Monk and Dr. Albert Mendeloff, of Johns Hopkins University, compared the emotional and personality traits of IBD patients with those of a healthy control group and found no significant difference. The consensus among gastroenterologists holds that even if psychological pressures contribute to IBD, something else must be fomenting the disease.

Although doctors may differ over what prompts the condition, they are unanimous in asserting that patients would be much better off if they were diagnosed earlier. Because

A technician prepares a formula for babies with the abnormally short intestines of short-gut syndrome. The infants are fed successively larger and more concentrated doses of the formula, developed by Dr. Dennis Shermeta at The Johns Hopkins Hospital. The formula stimulates the food-absorbing intestinal villi to grow denser, increasing their capacity. In two to 12 months, such children—who otherwise would be condemned to a life of intravenous feedings—can eat normal foods.

symptoms may mimic those of lesser ailments such as irritable bowel syndrome *(Chapter 3)*, many victims harbor a wishful hope that the problem will clear up by itself. Conversely, the sight of blood in the stool may be so frightening that the patient avoids consultation, fearful of what the doctor may find. Certainly anyone experiencing diarrhea for more than a few days or signs of blood at any time should see a physician. Pain and fever call for even more prompt inquiry.

Similarity between Crohn's disease and ulcerative colitis makes them hard to differentiate, but doctors strive to distinguish between the two disorders in each patient. The diseases tend to develop along different lines and thus sometimes call for different therapies.

A physician suspecting IBD will want to inspect the lower section of the colon by means of a sigmoidoscope, a type of endoscope that permits detailed examination of the mucosa. If the patient has ulcerative colitis, this will reveal it. Crohn's disease is more difficult to pin down, however. A full GI series of X-rays will usually be required—the stomach and small intestine being outlined by means of a meal containing a substance opaque to X-rays, and the large intestine photographed after the patient has been given an X-ray-opaque enema. Skilled radiologists can usually differentiate ulcerative colitis from Crohn's disease just by looking at the films: Crohn's shows a distinctive thickening of the intestinal wall; ulcerative colitis offers a thin, granular aspect.

To confirm the diagnosis and also to rule out the possibility of cancer, the physician may take an intestinal biopsy. The doctor may also ask the patient to drink a few glasses of milk to check for lactose intolerance—an inability to digest milk products. Sometimes this affliction rather than IBD causes the symptoms; even if it does not, an intolerance can make an IBD patient's diarrhea even worse than need be.

A bewildering variety of treatments that help

The cornerstone of treatment of IBD as practiced by most gastroenterologists is the careful orchestration of certain drugs. Since every patient's particular disorder is idiosyncratic and because no two persons will react to a drug the same way, physicians try by experimentation to arrive at a program of medication that will control the disease and minimize the effect of acute attacks without unpleasant side effects or other complications. Two classes of drugs are generally considered effective—the sulfas and the steroids; a third, the immunosuppressives, is more controversial.

The most widely used sulfa compound, sulfasalazine, a chemical relative of aspirin, seems to help about two thirds to three fourths of all ulcerative-colitis patients, but it is less effective against Crohn's disease. Sulfasalazine, like aspirin, appears to reduce inflammation by interfering with the body's production of prostaglandins, internal chemicals that sensitize tissues to inflammatory pain. The drug is especially useful in preventing relapses, but if taken in quantity it sometimes causes nausea, headaches and skin rashes.

Along with sulfasalazine, or as an alternative, the doctor may try one of the hormone compounds, such as cortisone or prednisone, that reduce inflammation. However, their side effects can be serious, making hair grow in the wrong places, and bringing on skin disorders, high blood pressure and diabetes. The physician generally prescribes the smallest feasible dosage—perhaps increasing it temporarily if a relapse occurs—and uses it only on a short-term basis. These drugs "buy you the time you need to allow things to improve," said New York gastroenterologist Nathaniel Cohen.

If either of these agents proves unavailing, the physician may resort to one of the immunosuppressive drugs, which alter the body's immune response. The best-known are azathioprine and a derivative called 6-MP. They are controversial, as they can cause a decrease in white blood cells that may lead to cancer of the lymph nodes. But a study at New York University School of Medicine and New York's Lenox Hill and Mount Sinai Hospitals indicated 6-MP to be "an effective and useful agent in the management of Crohn's disease." Said Dr. Cohen, "I would use it only as a last resort, as we don't yet know the long-term side effects."

In addition to drugs, the prescription includes a low-fiber diet—the ailment may have narrowed the intestine, threatening obstruction if indigestible fiber is introduced. For others, fibrous foods may or may not be beneficial—doctors tend to feel each patient will react differently.

Whatever the cause of their ailment, many IBD victims seem to benefit from psychiatric counseling. Followers of Dr. Sperling recommend an immediate psychiatric evaluation at the onset of the illness, and continued therapy thereafter, especially for a child. But most gastroenterologists, while acknowledging that time on the therapist's couch may help, warn that other treatment is invariably needed. Cautioned the National Foundation for Ileitis and Colitis, "In no case should any patient receiving psychotherapy be deprived of the more standard medical, and at times surgical, care."

The question of surgery is always in the background, in fact, with IBD. An operation may be recommended if the patient's condition, despite every effort, continues to worsen, if the disease becomes life-threatening or if a malignancy is found. Because Crohn's disease is often severe, almost half of all patients are eventually operated on—despite the fact that surgery is no cure and the ailment is fairly certain to recur. Ulcerative colitis, which is confined to the colon and rarely spreads, is often cured by surgery, but operations are needed in only about 10 per cent of the cases.

Thoughtful physicians meanwhile are all too aware, not only of how much is known about IBD but of how much is not. "Recently," said Dr. Sachar, "we and other investigators discovered in examining the intestinal tracts of Crohn's patients that the unaffected parts of their colons—the healthy, not the diseased sections—had tiny abnormalities in them: microscopic holes, much too small to be seen with the eye, but definitely there." What the holes mean, whether they provide a toe hold for bacteria, and whether they lead to future infection, Dr. Sachar could not tell.

In the face of such uncertainty, doctors try to keep an open mind while hoping for a discovery that will lead to a cure. In one article, Dr. Sachar and two associates speculated that IBD's cause could at any time "be stumbled across serendipitously, perhaps by a veterinary pathologist, food technologist, dental technician, or water pollution control engineer." For the moment, the paper went on, "the only practical approach appears to be to keep working in good faith along all traditional pathways, while keeping our minds receptive to unconventional ideas." ❋

The woman who eats as she sleeps

Drifting off to sleep, the woman at right is also lying down to her meals: Before dawn she will have consumed the equivalent of breakfast, lunch and dinner without ever opening her mouth. Her body is fed intravenously by the pumping device beside her bed, which sends liquid nutrients directly into her bloodstream to bypass a digestive tract so damaged by disease it is almost useless.

This radical method of ingesting daily fare, called total parenteral nutrition, not only prevented Lee Koonin from wasting away, but made possible a life that in most other respects is a normal, active one. The 45-year-old Massachusetts woman, afflicted with Crohn's disease, a chronic malady in which deep ulcers may attack at any point along the 30-foot digestive canal, had undergone 19 separate operations beginning shortly after the birth of her first child. Over two decades doctors removed all of her colon and rectum and most of her stomach and small intestine. She was left with a three-foot digestive tract ending in an ileostomy—an opening that passes waste from the end of her small intestine out into a bag to be emptied during the day.

With so little of the tract remaining, Lee Koonin could not eat enough in a normal way to keep her from the edge of starvation. Food sped through her like water through a sluice; only a fraction of the nutrients were absorbed, and she lost weight steadily—from 112 to 65 pounds. "The old expression 'You are what you eat' is not accurate," observed her husband, Marshall. "You are what you absorb."

Lee Koonin was virtually starving to death, suffering memory loss and epileptic-like seizures, when a Boston doctor made a startling proposal: "I would like you to think about never eating again." Her first 10 days on parenteral nutrition brought a dramatic 10-pound gain. Her alertness and energy returned, enabling her to pick up long-abandoned activities—tennis, music, sailing—that she came to value more than ever. "Except for mountain climbing," she said, "there really isn't much that I don't do."

Lee Koonin rests in bed after hooking up her nightly quart of fluid nutrients to the machine (foreground) that pumps it into her bloodstream as she sleeps. The plastic tubing that runs across the pillow is taped to her shoulder to keep it secure when she moves during the night. The end of the tube attaches to a catheter entering her body just below her breastbone.

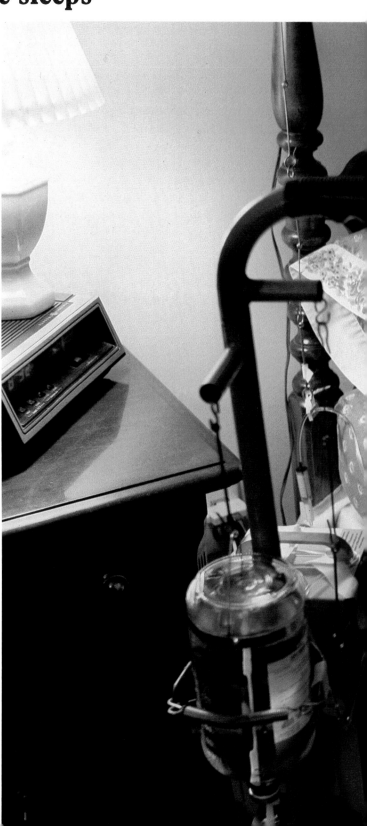

In the living room of their home near Boston, the Koonins sort out a delivery of liquid nutrients and other supplies—the 33 cases will feed her for a month. Her personal pharmacy, where she stores and mixes the materials, is behind her in the lighted alcove.

In her pharmacy, Lee Koonin injects
minerals and vitamins into the bottle that
nourishes her. To a base providing 1.4
ounces of protein, she adds enough of the
sugar dextrose to make up 1,000
calories. She gets another 300 to 600
calories from normal eating—her system
can still absorb the sugars in food.

Stored in the refrigerator with the
family's food, the nutrient solution will be
taken out two hours before hookup time
to warm to room temperature. A single bottle
costs $50—one premixed at a pharmacy
would be at least $25 more—but part of the
expense is covered by government aid.

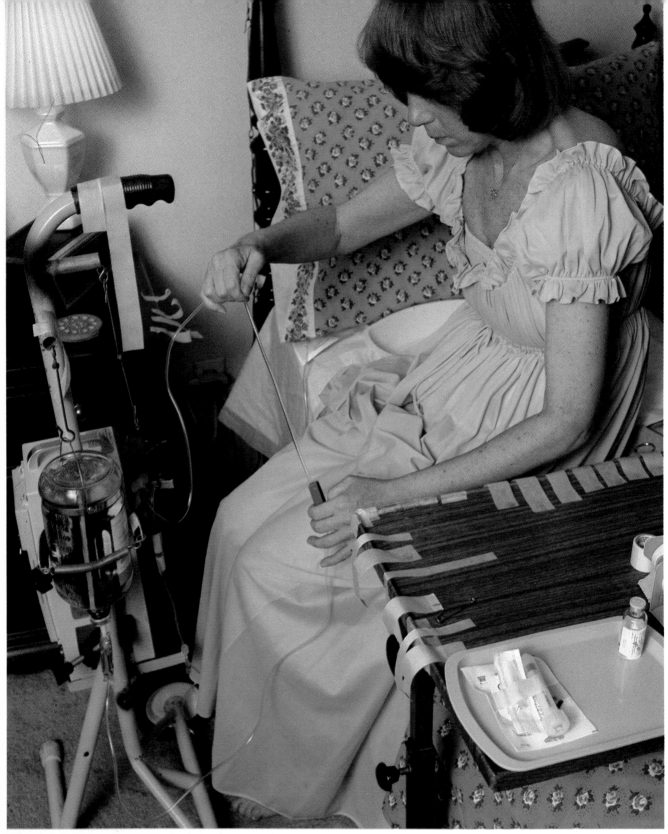

Tape and other materials on a tray beside her, Lee Koonin straightens out fresh plastic tubing for her nightly hookup. The battery-operated pump, equipped with a device that beeps if something goes wrong, is mounted on a rolling cart—''the best accessory I own''—which enables her to move around the house while connected to the pump.

Lee Koonin, her nightly infusion under way, practices on the living room piano: The feeding tube is tucked in under the neck of her robe, and the pump cart has been brought downstairs. Marshall listens in the background to her playing, which grew rusty—along with other coordination skills—during the years when her brain was starved for nutrients.

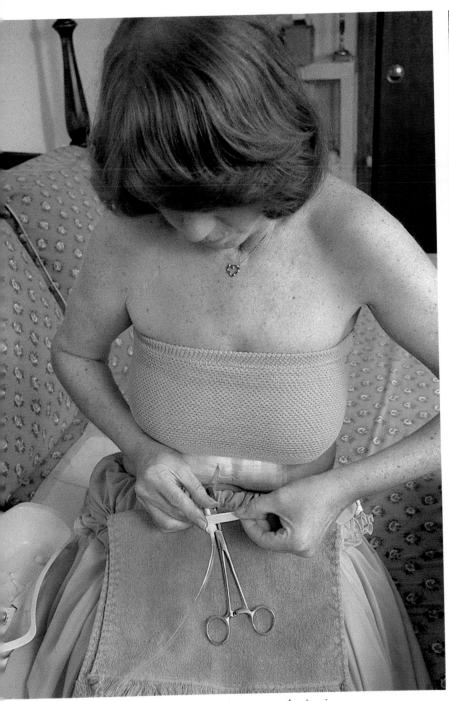

To make the connection, a tube from the pump mechanism is taped to the outer end of a catheter that is permanently inserted in the body and capped during the day. The inner end of the catheter joins a vein leading to the heart—so before it is uncapped for hookup, the catheter must be pinched shut with a scissors-like surgical clamp to keep blood from flowing out.

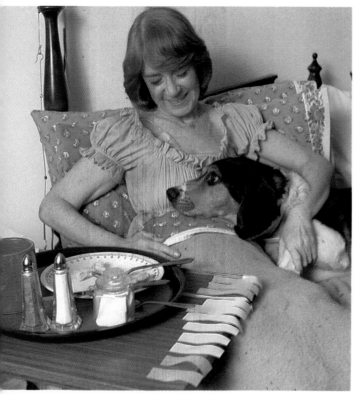

In the morning, after breakfasting in bed on a few nibbles of egg and toast—more for pleasure than for nourishment—Lee Koonin rewards the family dog, Nicholas, with leftovers. Next she will unhook her feeding apparatus, clean the catheter with an anticlotting chemical, recap the end and tape it securely under a bandage before beginning the day's activities.

Dwarfed by her tall boating instructor, Bob Gaffney, Lee Koonin sails on nearby Massapoag Pond with Marshall and their younger child, David. Four years of parenteral nutrition restored this victim of Crohn's disease to her normal weight—and by allowing her digestive tract to rest, kept her free from troublesome flare-ups of the affliction.

CHAPTER 6

The surgeon's masterly skills

Is this operation necessary?
Why a second opinion pays
The art and science of removing a gall bladder
When healthy appendixes must be cut out
For ulcers, meticulous snipping of nerves
The benefits and risks of intestinal surgery
Supreme audacity and skill—the kidney transplant

When a gunman's bullet smashed into Pope John Paul II's abdomen and through his intestines in the spring of 1981, there were widespread fears that he would not survive. Aside from the shock to his system, the tearing of his digestive tract allowed intestinal bacteria—harmless when confined to the colon—to escape into the abdominal cavity, threatening lethal infection. That he not only survived but made a swift recovery was due partly to the skill of his surgeons, partly to the modern techniques of gastrointestinal surgery.

The treatment of the Pope provides a dramatic illustration of the complex repairs that, once undreamed of, are today almost routine for the digestive system. The Pontiff's doctors found that his small intestine had been torn in five places and his sigmoid colon badly ruptured. The situation was grave, for the cavity was filled with blood and body wastes.

After clamping the severed arteries and veins to stop the bleeding, the surgeons removed the accumulated wastes with a suction device. To repair the badly damaged small intestine, they cut out two sections, one 14 inches long and the other 16, and sewed the cut ends together; the Pope could function well with this much loss. The torn colon was less seriously injured, and they simply stitched it together. But while the intestinal tract was now intact, there was a danger that the wound in the colon might be infected by feces once the patient began digesting food again. Therefore, the doctors provided the Pontiff with what has become a mainstay of abdominal surgery. They made a temporary colostomy—an opening in the abdominal wall through which wastes could

be drawn off while the wound healed; it could be closed later after recovery. The Pope had his colostomy closed three months after the shooting. His bowel was back to normal.

Aiding in his recovery were a number of substances that have revolutionized digestive surgery. Antibiotics are almost taken for granted nowadays; administered to the Pope, they prevented bacteria from multiplying and helped his body fight infection. Nutrients and chemicals, introduced directly into his bloodstream, nourished his body during several days of enforced fasting immediately after the operation. Anticoagulants combined with mild exercises—His Holiness was instructed to wiggle his feet every few minutes—staved off the formation of potentially dangerous blood clots. And the blood component gamma globulin protected him against the hepatitis that, despite all precautions, is always a threat from blood transfusions—10 pints were given the Pope during and after his surgery.

The Pontiff was 60 years old at the time of the shooting. One observer noted that if he had been similarly wounded when he was a young man, given the state of medical science then, he would almost certainly have died within days or even hours. Instead, after only 21 days he emerged from the clinic walking unaided and smiling.

Similar triumphs are achieved almost every day in any large hospital anywhere in the world, more often to treat digestive illness than gunshot wounds. A study of some 9,000 major operations revealed that mortality rates for digestive- and genital-tract surgery had dropped 60 per cent

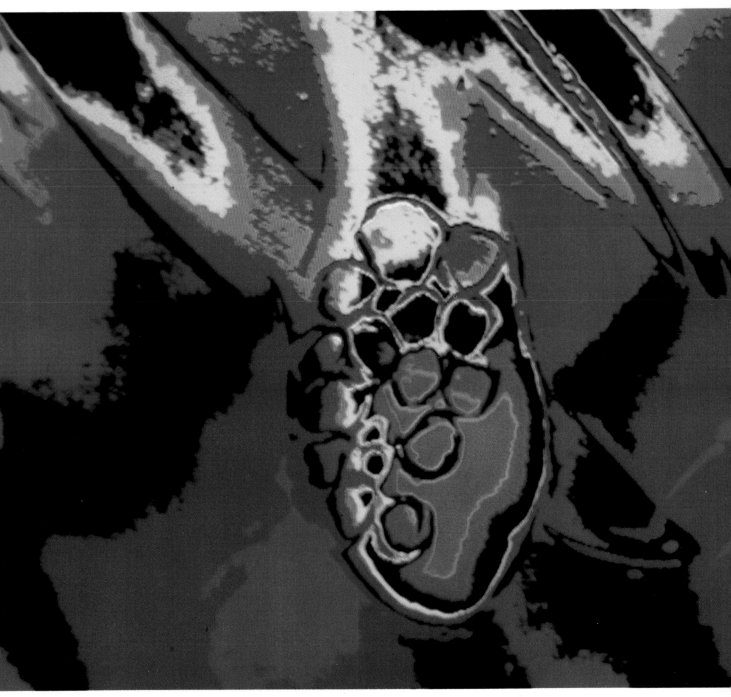

In this computerized X-ray, enhanced with colors to contrast tissues of different densities, thin yellow outlines trace the ovoid shape of a gall bladder, clogged near the top with the dark, round forms of gallstones. Capable of producing severe pain, the stones frequently occasion the most common of all digestive operations—removal of the gall bladder.

over three decades. During a single 10-year period New York's Columbia Presbyterian Medical Center reported a decline of 80 per cent in mortality from stomach surgery.

More important, surgical safety has improved during a time when surgical techniques have grown ever more audacious and refined. Surgeons can now repair almost any section of the system or, using techniques like those employed for the Pope, they can cut away large sections and bypass them. Most dramatic of all, they can replace a failed kidney system with a transplant, a lifesaving operation that is performed on 4,000 patients in the United States every year.

Physicians are familiar with the ways the body usually reacts to surgery, and they take endless precautions against error. Almost always, there are no hitches. But two facts make some hitches unavoidable. One is that surgery is an inexact science, hedged about by uncertainty. The great Canadian physician Sir William Osler once told a class of medical students that half of what they had learned was probably wrong—then added that, unfortunately, he could not tell them which half. Secondly, almost every operation involves a trade-off—for everything gained, something is lost. A stomach or intestine, for example, may be restored or repaired, but the system is never precisely the same again.

Is this operation necessary?

Except for emergencies like the one that confronted the Pope, most surgery involving the digestive tract aims to relieve chronic illness. It is elective—and thus almost inevitably debatable. Overall, digestive surgery is on the wane, but debate continues to rage over the operations still being performed. On one side, supporters of surgery point out that it is frequently the sole remedy available; on the other, critics argue that advances in medication and other noninvasive treatments have meant that surgical procedures once taken for granted may now be unjustified.

Of course, any surgery remains a serious matter, never to be undertaken lightly. "When a surgeon cuts into a patient he is tampering with an extremely delicate physiological balance," said Dr. Myron Denney of New York City, "a complex system of intricate bodily functions only partially under-

stood by even the most knowledgeable scientists." When the balance is upset, the body may mobilize its defenses in ways that cannot be foreseen, producing drastic changes in blood pressure, breathing, even cell activity.

The debaters do have important points of agreement. Both sides recognize that surgery is almost always the only remedy for any one of four conditions that, if unattended, would threaten life. One is perforation—a break in the wall of the digestive tract that permits food or its digestive remnants to escape into the abdominal cavity. A second is hemorrhage—internal bleeding that refuses to stop. A third is obstruction—a blockage of the intestinal tract or its related ducts that cannot be released by medicines. The fourth is the existence or likelihood of cancer.

A number of other situations, however, can also call for surgery. Any infection or inflammation that does not respond to medication may require an operation; a gall bladder can become so inflamed that it must be removed, for example; or the toxic conditions produced by inflammatory bowel disease may call for swift surgical action. Occasionally, an operation must be performed for diagnostic purposes: If tests have failed to identify the source of an intestinal pain or disability, there may be no recourse but to enter the body for a definitive exploratory look.

One last general reason for surgery, particularly common in digestive diseases, is intractability: The ailment does not respond adequately to medical treatment, and the patient fairly begs for a deliverance that surgery alone seems able to provide. Dr. M. Michael Eisenberg of New York recalled the case of a 47-year-old secretary who suffered intermittently from an ulcer for 25 years. Intensive medication relieved the symptoms of each attack, but as time passed the attacks became longer and more frequent, the drugs less effective. She became depressed and easily fatigued. At length she appealed to her doctor for some new kind of relief. Surgery seemed the best answer—in her case, removal of part of her stomach. Her discomfort largely disappeared after the operation, and she was content with her decision.

In most cases of digestive disease, the best answer is likely to be the result of considered judgment by a medical doctor,

by a surgeon to whom he has referred the case—or, preferably, by both. But doctor and surgeon sometimes disagree.

In the ranks of medicine, no greater gulf exists than the one that separates surgeons from the many medical specialists who generally do no surgery. Whenever there is doubt in a case, surgeons, by training and inclination, are prone to cut; the knife, they contend, cures swiftly and efficiently. Other physicians view digestive surgery as a last resort, to be turned to only when all other techniques have proved ineffective; to them, it is a confession of failure. The gulf dates back many centuries, to the days when surgery was performed by barbers. Until the 19th Century, when anesthesia, antiseptic operating conditions and X-rays ushered in modern surgical methods, surgeons were looked down upon by medical men.

The schism still exists. Commented surgeon-author William A. Nolen in *The Making of a Surgeon,* "Medical men regard surgeons as technicians: not too bright, but show them what has to be done and they may have the dexterity to do it. Surgeons look upon medical men as doctors who lack decisiveness. Internists hem and haw for hours over whether to give a patient penicillin or aureomycin; they'd be lost if they had to make up their minds in minutes whether or not to open an abdomen." The conflict is compounded by a third group of doctors—pathologists, the medical scientists who study the causes, development and consequences of diseases, along with their remedies. According to a medical saying, "Internists know everything but do nothing; surgeons know nothing but do everything; pathologists know everything and do everything, but too late."

One influence on surgeons' decisions about operating is the manner in which they conduct their practices. A surgeon must keep operating in order to maintain and improve his skills, particularly in complex and difficult procedures; the more colostomies he performs, for example, the more adept he becomes. And just as the best surgeons are usually the busiest, because they are most in demand, so the busiest ones are likely to be the best, because they do not allow their skills to get rusty. It is not only their own skills that must be maintained but also the skills and coordination of their operating teams. Much modern surgery is so complex that it requires a large group of specialists practiced in working together. A kidney transplant, for example, calls for 14 trained experts: two chief surgeons, two assistant surgeons, four anesthesiologists, four scrub nurses and two additional nurses—all so familiar with one another's procedures and needs that every step goes almost automatically.

Another acknowledged influence on the surgeon's decision is, simply, money. Most surgeons get paid for operating, not for declining work. They live by their fees. Many physicians—and some surgeons—feel that the fee-for-service system produces a conflict of interest that can work against the patient. Said Dr. George Crile, Jr., of the Cleveland Clinic, "Training, practice, and economic pressures can push the surgeon's thinking in the direction of more frequent, more radical, and more remunerative surgery.''

At issue, time and again, is the matter of operations that are unnecessary. According to a Congressional study, in one year more than two million unnecessary operations were performed in the United States and resulted in some 10,000 needless deaths. Except for a handful of what journalist Lawrence Galton called ''outright charlatans performing needless operations simply for the fees,'' those responsible for these mistakes are well-intentioned doctors who have made decisions that, at the time, seemed well founded.

A number of defenses do exist against unneeded surgery. Most hospitals have so-called tissue committees, which analyze whatever is removed at every operation—an ulcerated stomach lining, for example—and reprimand any surgeon who in their opinion has operated ineptly or needlessly. Too many reprimands can result in the loss of a surgeon's right to operate at the hospital. The American College of Surgeons can also discipline its members. More important, the surgeon has his own reputation to protect, is mindful of malpractice challenges and is bound by the Hippocratic oath to ''abstain from whatever is deleterious and mischievous.''

Why a second opinion pays

Perhaps the best protective device against unwarranted surgery, however, is the second opinion, an independent review of the decision to operate, provided by the most highly qualified expert obtainable. A New York City labor union discovered that when it asked a second physician to review each recommendation of surgery for its members, the number of operations actually performed went down almost 20 per cent. Some health-insurance plans not only urge members to seek second opinions, but pay the cost; the extra expense is more than balanced by the savings on surgery.

Just what constitutes a useful second opinion can be elusive, however. Consider a case in which a family physician thinks a patient needs an operation and calls upon a surgeon for corroboration, and the surgeon agrees. This is not truly an independent second opinion; the patient has essentially received only one surgeon's view—and it is possible that the two doctors, having worked together before, tend to approach some medical problems the same way. What is more, a physician can, if he chooses, virtually direct the kind of surgical opinion he prefers. One gastroenterologist remarked, ''If I think surgery might be indicated but I'm not bullish about it, I will often refer the case to a friend of mine who is the most conservative surgeon in town. I can predict he's going to take a dim view of going ahead.''

The patient is far more likely to obtain a truly independent view by asking the family physician or the primary surgeon to recommend a third professional, who may see the problem differently. Ideally, it should be someone with no direct ties to either of the two physicians. One possibility is a surgeon attached to a different hospital, though such authorities as Dr. Myron Denney have argued that the safest course is to consult not another surgeon but a specialist in the area of concern. In any case, the patient need not rely entirely upon his own doctor or doctors to get the names of additional experts. The *Directory of Medical Specialists,* available at most public libraries, provides the names of local authorities. Another source is the nearest medical school, and if the school maintains an outpatient clinic, so much the better; the clinic staff, recruited largely from the medical faculty, may contain some of the leading authorities in the country.

The primary physician still has a role to play; he should provide the findings of tests and X-rays to make the second opinion faster and less expensive to produce, and if the second opinion is in disagreement with the first, the patient should expect all the doctors involved to work out the conflict. A second way of resolving such a disagreement is to obtain still another opinion and accept the majority view. A third is for the patient to follow his own feelings and instincts in deciding upon the course of treatment.

The third alternative is not as surprising or foolhardy as it may seem. In the end, the real object of a second opinion is to help the patient make up his own mind, for the final decision is the patient's. Said Dr. William R. Barclay, an official of the American Medical Association: ''Patients should realize that they're the boss, since they are purchasing a service.'' Certainly, competent surgeons will not object to outside advice, for they, more than anyone else, realize that many surgical questions are iffy, and many operations questionable.

One operation that may increasingly be questioned is gall bladder removal. It is now the most common of all digestive operations, performed on some 500,000 people annually in the United States alone, but the development of chenic acid, or CDCA, raises the possibility that all gallstones may someday be dissolvable with drugs. CDCA, however, works only on one kind of gallstone, and mainly for stones that are not very large. Other patients must get rid of stones by having the gall bladder removed. The operation itself is considered unremarkable, even routine, nowadays—yet it is major surgery for a potentially life-threatening ailment. The surgeon must work upon a relatively inaccessible organ nestled deep within the intricate folds of the small intestine; he must examine not only the gall bladder but any neighboring structure that may contain stones; and after removing the bladder he must make sure that the common bile duct, which links the bladder, the liver and the intestine, is free of toxic matter. It is a triumph of modern technique that the entire procedure, which generally takes an hour or more to perform, now has a mortality rate of only .3 per cent.

The art and science of removing a gall bladder

To understand how a record like this has been achieved, it is instructive to track an actual operation, such as one that occurred at Mount Sinai Hospital not long ago. The surgeon was Dr. Arthur Sicular, a specialist in surgery of the digestive tract. On the operating table that morning was a woman in her middle 30s. She had suffered for years from heartburn and digestive distress; more recently, she had had intermittently severe pain in the upper-right quadrant of her abdomen. X-rays revealed several stones, each about a quarter of an inch across, in her gall bladder. Her doctor recommended removing the gall bladder.

There was little need for a second opinion in this case. Sometimes a routine X-ray, taken for some unrelated purpose, may reveal gallstones that have not yet caused any symptoms, and doctors disagree about whether or not such "silent" stones should be removed. Some argue that, because there is a good chance that the stones will eventually

become troublesome it makes sense to get rid of them, particularly if the patient is under 40. Others contend that the possibility of future trouble is too slight to warrant surgery, and that, particularly for the patient over the age of 55 or 60, the risks—leakage of bile, infection, injury to the ducts—tip the scales against proceeding. This patient did not raise that doctor's dilemma: She was young, she had felt severe pain, she almost personified the prototypical gallstone victim, and the decision was properly made.

The operation began with the time-honored ritual of scrubbing, in which all those who will touch either the patient or the surgical instruments wash their hands and forearms for a prescribed time, usually 10 minutes—because it takes that long for the scrubbing action and the antibacterial soap to have their cleansing effect. Dr. Sicular then walked into the tiled operating room with his arms extended, his nose and mouth masked and his hair covered. He was gowned and gloved by a nurse. Similarly garbed were a surgical assistant, a resident in the last stages of surgical training, who would help with the operation; a medical student, who would both observe the operation and take part in it; and the so-called scrub nurse, who would hand the surgeon his instruments.

By now the patient was on the table, unconscious and with her eyelids taped shut (if they were to open while she was anesthetized, they might dry out, as she would not blink normally); the resident was coating her abdomen with a reddish-brown antiseptic. Also present and masked, but not "in scrub," were two supply nurses, ready to provide whatever was needed, and an anesthesiologist and his assistant, bending over a bank of elaborate monitoring devices.

In this operation, as in all major surgery, the role of the anesthesiologist would be crucial. He was not merely an anesthetist, a trained technician who administers anesthetics under a doctor's direction, but a doctor in his own right. Before the operation he had studied the patient's medical history, looking for special sensitivities or allergies to drugs, for reports of liver or kidney disease that could hamper the elimination of drugs, and for an existing pattern of medication that could cause dangerous drug interactions. During the operation, he would monitor not only the anesthetic dripping into his patient's blood, but such vital signs as temperature, blood pressure and heart activity; if necessary, he would sustain or strengthen these signs with transfusions of blood, blood plasma and glucose, and with additional drugs. Later, he would be on hand to keep the patient comfortable and well during the critical hours after the operation.

For this operation, the preliminary steps were almost completed. Soft piped-in music filled the room as with one gloved finger Dr. Sicular traced the line of incision—straight down for about three inches from a point just below the ribs, then diagonally down to the right side for another four inches or so. He asked for a scalpel. It was 8:34 a.m.

With surpassing gentleness—in a move described by one surgeon as "a long, deliberate tickle"—the doctor made his first cut, barely through the skin. Then slowly, carefully, he

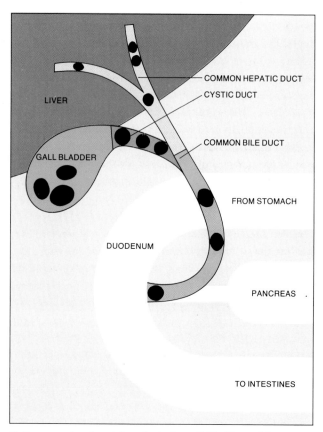

Gallstones—hardened deposits in the gall bladder and liver, or in ducts that lead from those organs—can obstruct the flow of bile, the greenish chemical that helps digest fats. Most gallstones are harmless, but if the stones block any ducts carrying bile or pancreatic juices to the duodenum, the bile backs up, causing sharp pain and, often, the yellowish skin of jaundice.

made successive cuts down through the layers of fat and muscle until he reached the peritoneum, the whitish, translucent membrane surrounding the abdominal cavity.

Certain features of the operation would already have been noticeable to a visitor. One was the lack of blood. Whenever Dr. Sicular sliced through a small blood vessel, the assistant clamped it to stop the flow; then, using a high-temperature electric cauterizer, which causes blood to coagulate instantly, he touched the end of the vessel, closing it—a small hissing could be heard—and released the clamp. Any remnants of blood were swiftly blotted up with sponges, leaving the field clear for the surgeon. Larger amounts of blood were whisked out by a suction device like a small vacuum cleaner.

Perhaps more striking still was the business-like but unhurried air of the entire proceeding. Modern anesthetic methods, besides sparing the patient agony, remove what once was one of the cardinal hazards of surgery—the need to rush.

There was no pausing, however. At 8:42, after setting the abdominal wall out of the way by stitching its edges to surrounding green towels, the surgeon punctured the peritoneum; for this part of the job he nicked the membrane with a scalpel, then used a pair of blunt-tipped surgical scissors, rather than a scalpel, to cut it open, to avoid injuring the tissues beneath the membrane. Anyone close enough might have heard a tiny rush of air, for the abdominal cavity is normally airtight; after the operation, the body would absorb the air now entering. After lengthening the cut, the surgeon reached into the cavity and maneuvered the intestinal loops aside so that he could get to the gall bladder. To hold the intestines out of the way Dr. Sicular carefully packed them back with two soft cloths moistened with saline solution. At 9:05 he pointed to a tube about as thick as a pencil. "The common duct!" he announced to the student.

Nearby was the target of the operation—the gall bladder and its pain-producing stones. Dr. Sicular did not immediately remove it. Instead, with great precision, he cut and pried away its surrounding tissue, leaving it attached to its ducts, then he lifted it away with a scissors-like clamp and rested it at the edge of the wound. "Okay," he called out, "we're ready for the X-ray." There was no need to X-ray the gall bladder—it would be coming out. The focus was on the common duct, which might be hiding stones that had eluded previous X-rays because they were too small or because the dye used to mark them had not reached them. If any were present, Dr. Sicular would have to get them today; a blocked common duct can cause liver damage. A technician wheeled an X-ray machine into place, the entire surgical team retreated to an anteroom, and the X-ray clicked off its shots.

While waiting for the X-ray film to be developed, Dr. Sicular had his assistant take a liver biopsy. Tests the previous day had shown the patient's liver to have slightly elevated levels of enzymes, a possible indication of abnormality or even of cancer. The assistant, guided by Dr. Sicular, thrust a long hollow needle, fitted to a plunger device, into the liver. He pulled back on the plunger, withdrawing the needle and with it a tiny section of liver tissue. "Excellent," came the Sicular verdict on his performance. (Happily, the biopsy proved negative—there was no damage to the liver.)

At 9:25 the developed X-rays were brought in and clipped to a viewer. "Good," said the doctor. "No obstruction. Everything clear as a bell." He could leave the common duct as it was. Immediately he proceeded to remove the gall bladder; in a few seconds he cut its connecting ducts—the assistant tied them off quickly to prevent bile from dripping into the abdominal cavity—then placed the separated organ in a metal dish on the instrument tray. Pink and shriveled, about three inches long, it hardly looked like the kind of thing that would cause so much pain and concern.

"Let's just take a quick look around now," said Dr. Sicular. The principal job done, the patient would get a bonus: an expert inspection of her abdomen. The surgeon reached in, felt around—with his years of experience he could judge an organ's health simply by feeling it—and began calling off his report. "Stomach's fine. Transverse colon seems good. Spleen okay." He found a minor diverticulum, or pouch, on the descending colon—nothing to worry about.

Then he reached up toward the chest. "Yup, just as I thought. She's got a real hiatal hernia." A defective, enlarged opening in her diaphragm had allowed the top of her stomach to ride up into the chest cavity; almost every day

gastric acid had seeped into the esophagus—the cause of her heartburn pain. "Here, see what I mean," said Dr. Sicular, withdrawing his hand and nodding to the assistant. The resident reached in and said yes, he could feel it: a clear gap between the diaphragm opening and the stomach top. "Now you," said Dr. Sicular to the student. She stretched to get her arm into position and thrust it in—then shook her head. "I can't find it." Dr. Sicular gently guided her forearm in the right direction. A few seconds passed. Then her face brightened. "Oh, yeah," she said.

The operation would not be extended to include a repair of the hernia. Surgery is generally reserved for those who feel unremitting pain—this patient had not—and it frequently fails to eliminate distress. Most doctors feel that heartburn can be controlled more effectively through medication and a change of dietary habits. Dr. Sicular's on-the-spot affirmation, aside from its immediate benefits to the medical education of the assistant and the student, was principally of interest to the patient and her doctor; there was now tangible evidence of what had been causing part of her distress.

At 9:37 a.m., just over an hour after the first incision, Dr. Sicular placed two drains—soft rubber tubes, each about eight inches long—at the corner of the wound, with their ends at the gall bladder site. They would be left in position for several days to carry off any remnants of bile. Quickly but carefully, he pulled out the packing cloths and restored the intestines to their original positions. Only a minute later, at 9:38, he called, "Ready to close."

As the doctor held the edges of the peritoneum free, his assistant grasped a curved needle with a long forceps; from one end of the needle dangled a foot or so of blue plastic suturing thread. With a deft turn of his wrist, the assistant passed the needle through an edge of the peritoneum and pulled the thread through. As the surgeon held the opposing edge steady, the resident pierced it, too, then brought the two edges tight together. Dr. Sicular tied two or three knots and the medical student snipped off the excess thread. The procedure was repeated along the full length of the incision, at points an inch or so apart. In the past, dissolvable sutures, frequently catgut, were generally used inside the body; today, many surgeons prefer plastic because it is less likely to get infected. It does not, however, disappear. Within her body, the patient would carry those blue knots—fashioned by Dr. Sicular, tugged tight by his assistant and trimmed by the student—for the rest of her days.

With the peritoneum closed, Dr. Sicular turned over the final stitching to his colleagues; the visible row of external sutures that his patient would proudly exhibit to friends and relatives as evidence of her ordeal was, like that of most operations, the work of a subordinate. Dr. Sicular turned to leave the operating room, then paused. Picking up a scalpel, he headed for the metal dish containing the gall bladder, sliced open the bladder and fished out three gallstones— each a quarter of an inch across, exactly as the X-rays had indicated, and each hard as a rock. He gave one to a nurse and told her to save it for the patient.

Why healthy appendixes must be cut out

With its elaborate procedures and niceties, this routine operation had lasted well over an hour. Surgeons generally consider gall bladder removal about midway in the range from the least to the most difficult digestive-tract operations. Appendectomies and hemorrhoid removals, in which relatively accessible parts are cut out of the body, are easier and simpler. More complex, and calling for some of the most sophisticated techniques of modern surgery, are ulcer operations, the excision and repair of intestines damaged by inflammatory bowel disease, and kidney transplants.

The easiest and most common of all these operations, the appendectomy, is so straightforward—the organ lies near the surface, and there are no surgical alternatives to merely going in and getting it—that it is a standard training procedure for students and residents. In the absence of complications such as perforation, the mortality rate is less than a tenth of 1 per cent. The appendix seems superfluous to body function and will not be missed. Patients are generally out of bed the following day and home a couple of days later.

The question of whether to undertake this procedure is often seen as beyond debate. The patient feels stabbing pain in the lower-right quadrant, then a small degree of nausea;

the doctor finds an elevated white-blood-cell count, indicating an infection, and presto! the operating room doors swing open to admit the afflicted. The tiny, wormlike appendix is severed from the large intestine as swiftly as possible—on the assumption that an inflamed organ may perforate and infect the abdomen. There seems no time to lose.

As it happens, this classic drama is being played out less often today than in the past. Though appendectomies still rank second only to gall bladder removals in the number of digestive-tract operations performed, cases of acute appendicitis and the number of operations have declined steadily since the mid-1960s. Between 1966 and 1978, for example, the drop amounted to about 15 per cent.

One reason for the decline may be that many past cases were misdiagnosed. Physicians have discovered that the pain suggesting appendicitis may really be caused by such ailments as diverticulitis, Crohn's disease, acute infectious diarrhea or irritable bowel syndrome. Today, whenever they safely can, doctors delay operating and double check their diagnosis. They wait to see whether the pain moves elsewhere. They question the patient closely to learn about other ailments or symptoms; for example, if diarrhea develops, a doctor would be particularly suspicious—most victims of appendicitis tend to become constipated.

But even now as many as 25 per cent of all appendectomies involve the removal of an organ that proves to be normal after all. Strictly speaking, these operations are unnecessary, and a tissue committee might raise an eyebrow, but many doctors stoutly defend the excess. ''The conscientious surgeon,'' said Dr. Howard Spiro of Yale University, ''must remove a certain proportion of normal appendices if he is to avoid missing the one that has ruptured and yet has given rise to few signs.'' Similarly, surgeons defend the fact that appendectomies are by far the most common abdominal operations performed on children. Said Dr. Nolen, ''If a child has the signs and symptoms of appendicitis, if I think the child has appendicitis, I have an obligation to operate. If I don't and he has appendicitis, the appendix may rupture. A child can die of a ruptured appendix. So if I remove a normal appendix and if the child's symptoms were caused by indigestion, I've done an unnecessary operation, but not an unjustifiable one.''

In some instances surgeons remove an appendix that they know is perfectly healthy. Some recommend the procedure for patients planning long stays in remote regions where the quality of medical care is questionable. And others remove a healthy appendix during the course of an operation for another ailment. The reasoning behind such a decision can be persuasive: Seven per cent of all persons get appendicitis sooner or later; it makes sense therefore, while the abdomen is open, to make sure that this patient will never be among the 7 per cent. But the practice is controversial. Some surgeons now argue that they should never remove even the most insignificant and superfluous organ unless its removal is absolutely necessary. And their reluctance contributes to the decrease in the number of appendectomies every year.

New ways to eliminate hemorrhoids

Also declining are operations for removal of hemorrhoids, the dilated varicose veins at the lower end of the digestive tract that afflict 35 per cent of the population. The rate of operations has dropped even more sharply than that of appendectomies: In the years between 1966 and 1978, for example, it fell by about 40 per cent—and the reason is simple. Less drastic methods of treatment have been devised; today, the basic ailment is generally amenable to such medical treatments as changes in diet and daily habits *(Chapter 3)*. Only when sufferers are unable to control the enlarged veins is surgery called for.

In past decades, hemorrhoid-surgery patients were hospitalized for several days and might miss weeks of work. The procedure itself—cutting off the hemorrhoids with a scalpel—was very painful. Some surgeons still use this technique, particularly in severe cases. But most hemorrhoids can now be excised quickly and almost painlessly in a doctor's office. The commonest technique is to tie a tiny rubber band around the neck of the hemorrhoidal growth, cutting off its circulation; in a few days the growth withers and sloughs away. Another method uses extreme cold rather than constriction. A probe chilled to a temperature as low as $-256°$ F. freezes the growth; the blood within the distended veins turns

solid and clots; and the hemorrhoidal tissue soon disintegrates. This procedure, too, can be done in a physician's office, and it is equally fast.

For ulcers: a meticulous snipping of nerves

Far more serious and difficult—and controversial—is surgery for ulcers. Once it was very common—at the end of the 1960s, for example, the rate of peptic-ulcer operations was double what it was toward the beginning of the 1980s. The decline in incidence of the disease, coupled with better understanding of its causes and treatment, made operations less necessary. New drugs such as cimetidine and sucralfate, which relieve symptoms and promote healing, promise to accelerate the downward trend in surgery. But in the 19th Century, ulcer incidence was increasing rather than decreasing, and pioneering surgeons provided desperately needed cures. Some of the techniques they developed remain in use, alongside newer methods, for surgery is still needed in certain cases, and it offers virtually permanent relief.

In the simplest operation for ulcers, the surgeon stitches up the ulcerated holes in the lining of the stomach or duodenum. More common is surgery to decrease the flow of stomach acid. But the operation that paved the way for such surgery was drastic: It did away with part of the stomach. That trailblazing operation, performed for a cancer of the pyloric sphincter, was the work of the tall, genial Viennese physician, Dr. Theodor Billroth, who is generally considered the founder of modern abdominal surgery.

A great teacher, a surgical genius and, incidentally, a famous amateur musician (he loved to play chamber music with his friend Johannes Brahms), Dr. Billroth had already accomplished such feats as removing a part of the esophagus and sewing the severed ends together. For a time, his successful 1881 operation, in which he cut out a cancerous pyloric sphincter and joined the remaining part of the stomach to the duodenum, seemed a false dawn for modern surgery; his next two attempts were failures, and enraged Viennese citizens stoned him in the streets of the city. But within a decade Dr. Billroth had performed over 40 stomach removals with a mortality rate of less than 40 per cent—a dazzling achievement for his period, and an irrefutable demonstration that a patient could lose part of the stomach and survive.

In 1884 Dr. Billroth's technique was adapted for stomach ulcer by Dr. Ludwig Rydygier of Poland, who argued that the key to curing the disease was to reroute the stomach's contents away from the ulcer, allowing the sore to heal. Dr. Rydygier and later surgeons constructed a passage from a point partway down the stomach directly into the jejunum,

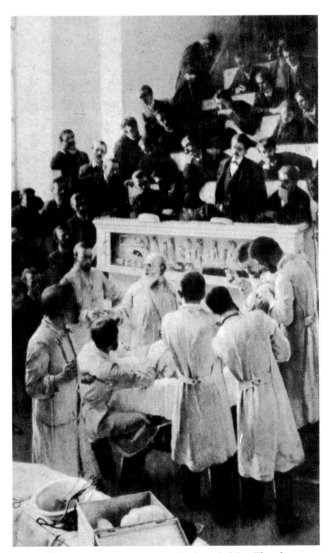

In this 19th Century illustration, white-bearded Dr. Theodor Billroth, the father of modern abdominal surgery, presides over an operation in the Allegemeine Krankenhaus, the hospital in Vienna where, in 1881, he performed the first successful removal of part of the stomach. The technique, still used in some ulcer and cancer cases, bears the famous surgeon's name.

the second section of the small intestine, bypassing the pyloric sphincter and the susceptible duodenum.

Sure enough, most ulcers thereupon healed—but others would soon appear in the bypass to replace them. Shortly after the turn of the 20th Century, physicians came to agree that it made more sense to ignore the location of the ulcer, and to remove instead the part of the stomach that secretes most acid. They assumed that the lower end was the villain, and this area became the target of surgery. As it happened, the assumption was false; the upper end of the stomach is the major secreter. But as luck would have it, the error made no difference, for the lower end secretes gastrin, the hormone that starts the acid flowing. Excising the lower part of the stomach, a procedure known as subtotal gastrectomy, became the standard ulcer operation. It cured more than 90 per cent of all patients, with a mortality rate of about 7 per cent.

That mortality rate was hardly ideal, of course, and doctors searched for better methods. A breakthrough came with the realization that the secretion of acid is controlled by the vagus nerves, which run from the brain through to the stomach. In 1943 Dr. Lester B. Dragstedt of the University of Chicago cut a patient's vagus nerves and the ulcer cleared up. Unfortunately the operation also inflicted a kind of paralysis upon the peristaltic movement of the stomach, slowing its emptying. To meet this problem, surgeons later devised refinements in which the drainage end of the stomach was enlarged. Eventually the two principal methods—the excision of the lower third of the stomach, and nerve-cutting— were combined in a single, relatively safe operation.

One side effect, however, consistently plagues patients after most ulcer operations: Their incomplete stomachs, far from emptying too slowly, move food along too quickly. This effect, called dumping, brings other symptoms in its wake—rapid heartbeat, sweating, mild nausea and light-headedness. The condition appears to be aggravated by foods rich in carbohydrates—cutting down on such delights as milkshakes, thick soups and rich desserts helps bring it to an end. Some sufferers are aided by eating smaller but more frequent meals, others by avoiding liquids during meals.

A refinement of the Dragstedt technique promises to eliminate the existing drawbacks. Called highly selective vagotomy, it consists of separating and severing just those fibers of the vagus nerves that control acid secretion; the other fibers, which control such functions as stomach-emptying, are left intact. The meticulous procedure cures the ulcer with few side effects, almost no risk—and little dumping. As in most modern ulcer operations, the mortality rate is near zero; the patient can expect a hospital stay of about two weeks and a normal life after another month or two.

The benefits and risks of intestinal surgery

For intestinal surgery, the picture is different. Mortality rates are higher, and life after surgery may be far from normal. But the value of the procedures and their constant advances in safety are indicated by their extraordinary increase in numbers, beginning about 1970. In the 1960s, fewer than 50,000 of these operations were performed annually in the United States; by 1972 the number had almost doubled.

Modern intestinal surgery is controlled not only by the nature of the disease to be cured (some 70 per cent of the operations are for cancer) but by the functions of the small intestine and the colon. The small intestine, which absorbs nutrients into the body, is the one truly essential part of the entire digestive tract; not much of it can be removed, and no one has invented a substitute for it. The colon, though it performs a useful service in absorbing water from the digestive residues, can be dispensed with, partly or completely.

The principal digestive ailments that may require intestinal surgery are diverticulitis and inflammatory bowel disease (IBD). All are regularly managed by drugs and diet *(pages 67-71)*, but an intestine sometimes deteriorates beyond the body's capacity for self-repair. When that happens, surgery may be the only remedy, its nature depending on the specific disease and the patient's condition.

Three major procedures are available. In one, the diseased part of the intestine is removed and the remaining parts are stitched together. In a second, part of the colon, usually including the rectum, is removed; body wastes exit through a colostomy—an opening the surgeon makes in the abdominal wall—and are collected in a small, disposable bag. In the third, the entire colon is removed, sometimes along with the very end of the ileum; wastes are brought out through a similar opening, called an ileostomy, to a collection bag.

The first and simplest of these techniques is generally used for diverticulitis, but a temporary colostomy may turn the operation into a two-stage or even a three-stage affair. In a typical situation, a section of the colon develops a large infected pouch, and the surgeon fears that simply removing the infected section will leave a danger of infection or other complications. Therefore, in the first stage of the operation, he constructs a colostomy, temporarily bypassing the affected area and giving the infection a chance to heal itself.

In the second stage, performed several weeks later, he removes the damaged section; if he closes the colostomy at this point, the operation is complete in two stages. But if he believes that the newly spliced colon needs time to heal before going back to work, the surgeon will leave the colostomy in place. A couple of months later, in the third stage of the operation, he will close the colostomy and restore the colon to its regular functioning, virtually good as new. But no competent surgeon uses more stages than he must—indeed, depending on the patient's age and condition, he may gamble a little to reduce the number of stages—because each stage carries a 3-to-5 per cent risk of death, and a somewhat greater risk of complications.

For IBD patients, the situation is more complex and the surgical outlook depends largely on whether the specific ailment is ulcerative colitis or Crohn's disease. Ulcerative colitis, which affects the colon but not the small intestine, is characteristically limited in area and will not strike in more than one place. Moreover, it carries a special danger: A victim who suffers the disease for seven to 10 years is at an increased risk of developing cancer of the colon. Therefore, doctors are likely to recommend surgery early—and they can promise a total cure, since the disease will not reappear.

Crohn's disease, on the other hand, presents a dilemma. The chance of its becoming cancerous is very low—a fact that argues against surgery. But patients often develop fistulas—abnormal passages from inside the intestine to the abdominal cavity—which may demand surgery. The problem

Three ways to cut a nerve—and cure an ulcer

When drugs fail to relieve a serious ulcer, surgery often can. The safest and most common ulcer operation is a vagotomy—cutting the vagus nerve, which stimulates secretion of digestive juices. Severing part or all of the nerve prevents the stomach from over-producing acid, and thus allows an ulcer to heal.

The point where the nerve is cut is of vital importance, for the vagus affects digestive secretions of organs other than the stomach. The three basic ways of cutting the many branches of the vagus are shown below. The oldest, easiest and most common technique is also the most radical. In a so-called truncal vagotomy, the two trunks of the vagus nerve are cut so that the secretions of not only the stomach, but also the gall bladder, pancreas,

small intestine and part of the colon are reduced. The procedure cures ulcers in nine out of 10 patients. However, according to one study, half the cured patients experience unpleasant, if tolerable, side effects—diarrhea, nausea, vomiting and abdominal pain.

A more refined operation, almost as old and just as effective, is the selective vagotomy. Both vagus trunks are severed, but in this case they are cut below their two main branches—the hepatic and celiac. Since fewer organs are affected, fewer side effects occur.

To further reduce the risk of side effects, the highly selective vagotomy was developed in 1969. In this procedure, only those nerve branches that serve acid-secreting parts of the stomach are cut. Fewer than 10 per cent of patients suffer adverse reactions.

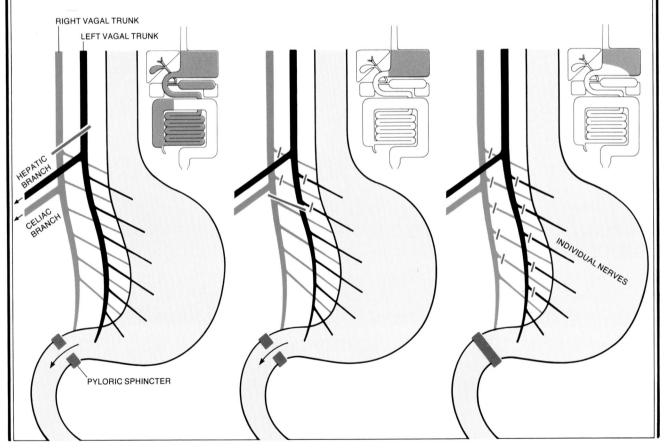

In a truncal vagotomy the vagus nerve is cut (red bar) above the celiac and hepatic branches, affecting the stomach and other digestive organs (shaded region, inset). A drain valve, the pyloric sphincter, is paralyzed, and must be widened surgically.

When the vagus nerve is cut below its celiac and hepatic branches, in a selective vagotomy, only the stomach (inset, shaded) is disconnected. As in a truncal vagotomy, the pyloric sphincter must be widened surgically to allow the stomach to drain.

In a highly selective vagotomy, an operation on tiny nerve branches, the vagal trunks are left intact. Only the parts of the stomach that secrete acid (inset, shaded) are affected. The pyloric sphincter is unimpaired and no drainage operation is needed.

is compounded by the fact that Crohn's disease is very likely to recur elsewhere after surgery—some estimates, in fact, set the recurrence rate as high as 90 per cent. With good reason, physicians are reluctant to advise surgery except in cases of absolute need, such as an intestinal obstruction, a perforation or a dangerous fistula. In some cases they may recommend a temporary colostomy or ileostomy, as for a diverticulitis patient, in the hope that the colon or the fistula will heal itself during a brief rest.

Obviously, IBD sufferers face a classic trade-off. Each must decide for himself whether the relief promised by surgery outweighs the inconvenience of an ostomy bag, the possible recurrence of Crohn's disease and the risks of the operation itself. The choice is difficult, but it is often dictated by sheer pain: After long periods of extensive and repeated pain, many IBD patients come to yearn for and even to demand virtually any remedy that will bring a semblance of normality back to their lives.

The patients whose colons are removed must accept the chores of installing and disposing of ostomy bags and cleaning the ostomy itself. Most difficult of all, perhaps, is resolving the emotional turmoil of replacing a peculiarly personal and intimate organ with an inanimate device attached to the outside of the body. One young Washington, D.C., woman recalled that when her doctor telephoned her with the news that she needed such surgery, she was so enraged that she threw the telephone across the room and screamed. Yet most people adjust well after the first shock. Their ordeal may be eased by counseling from volunteers who have experienced the same operation and are now leading healthy lives. In time, they discover that the disposable bags are odorless and almost totally undetectable beneath clothes. They can engage in most activities and virtually all sports, including swimming. A normal sex life is to be expected, and many have borne children.

The young woman who hurled the telephone was single at the time. Six months after her surgery, she met a man to whom she became deeply attached. She worried about how to tell him of her operation and how he would take it. Finally, one evening, she simply said, "Guess what I've got—an ileostomy." He replied that he had never heard of it, and asked her to explain. When she did, he remarked, "Why make a big deal?" Five months later they were married.

Supreme audacity and skill—the kidney transplant

The most daring operation on an organ associated with the digestive system—an operation now surprisingly common and successful—is replacement of the kidney. For the victim of severe kidney disease, the alternative to surgery is not discomfort or pain, but death—or a life of virtual slavery to a mechanical stand-in for the ailing organ *(pages 114-115)*. Although such a dialysis machine does a superb job of filtering the gallons of fluid that must be cleansed for the body every day, it does not perform such kidney functions as regulating the chemistry of the blood and balancing salt and water. It is not surprising that, according to a National Institutes of Health study, fewer than half the adults using the machines held even part-time jobs; by contrast, more than 90 per cent of recovered transplant patients worked full time.

The search for a surgical solution to the problem of hopelessly diseased kidneys was under way even as the techniques of dialysis were being perfected—and from the start, the search looked hopeful. Though the transplant of an organ from one human body to another called for a high degree of sophistication and audacity, the kidney made a particularly inviting target. Its major anatomical connections were few and fairly simple: an artery and a vein to carry blood into and out of the organ, a ureter to drain it of water and wastes. Because a human being can live with one kidney or even half a kidney, transplants could be taken from living donors. And if a transplanted kidney did not survive, the patient could be kept alive by dialysis until another attempt could be made.

Through the late 1940s and early 1950s, surgeons tried in vain to turn the promise of kidney transplants into a reality. Then, in September 1954, less than a decade after the first successful dialysis, a team of surgeons at Boston's Peter Bent Brigham Hospital cut a kidney out of the body of 23-year-old Ronald Herrick and successfully fitted it into the body of his twin brother, Richard. That triumph was followed by others: By the early 1980s, more than 40,000 successful operations

Flying kidneys to save lives

Every year about 1,000 human kidneys, each carefully kept alive at chill temperatures by a machine, are shipped for transplanting in the United States. Most travel on commercial airlines, strapped into passenger seats *(below)*. Removed in one city from a donor who has died suddenly, the kidney is flown to a distant hospital—perhaps thousands of miles away—where a hand-picked recipient and a surgical team await its arrival.

A portable perfusion device, about the size of an ice chest, preserves the extracted organ in a cool, sterile compartment and maintains it by flushing it constantly with fluids. A kidney can survive this way for about 72 hours. Knowing where to ship it depends on a network of computers linking almost all of the 150 or so American hospitals (and some in other countries) that perform kidney transplants. The computers store technical medical data about each of the more than 5,000 people who are awaiting new kidneys—information that helps doctors decide whether the immune system that defends the potential recipient's body would be likely to accept or reject the donated organ. When a kidney becomes available and a compatible recipient cannot be found in the donor's immediate area, a call goes out by computer for the names of others elsewhere who might benefit. The best match is chosen and, within hours, the life-giving kidney is airborne.

Secured by a seat belt alongside a technician courier, a kidney perfusion device preserves an extracted specimen—visible in the clear compartment at top left—en route to a distant city for transplant. All kidneys fly first class, and always in the first window seat, primarily for reasons of space.

had been done. The operation itself *(pages 146-163)* remains one of the most intricate in the surgical repertoire, but its procedures and its hazards have been mastered; in recent years mortality rates have ranged from 5 to 10 per cent.

For kidney-transplant patients the greater danger is not the surgeon's knife, but the patient's own body. In a process called rejection, the immune system of the body mobilizes armies of chemical and cellular agents to destroy the transplanted kidney as a foreign invader. The signs of its foreignness are patterns of surface chemicals called antigens, determined by the donor's genetic inheritance; the chemical defenders are called antibodies; and normally the antibodies would win out—the antigens on the transplanted tissue would not match those on the host's tissues, and the immune system would go into action to kill the new kidney.

Rejection did not occur in the case of Richard and Ronald Herrick because they were identical twins, with identical genetic make-ups; their tissue cells are indistinguishable. Richard's system was outfoxed—it did not recognize the new kidney's antigens as foreign. Five years were to pass before operations succeeded with genetically different subjects, and then success depended on using subjects with genetic make-ups that, though not identical, were very similar—siblings and parents.

The degree of similarity, and thus the chance of a good match between donors and recipients, depends upon the patterning of four different antigen groups found on all human cells. Collectively, they are called the HLA antigens; the groups are designated A, B, C and D. The likelihood of a perfect match between genetically dissimilar individuals can hardly be said to exist—it has been compared to the chance of hitting a jackpot on a four-window slot machine. What is more, the job of mapping the groups presents its own difficulties. Though HLA-A and HLA-B are easily identified in a blood test, the other two—particularly HLA-D, which plays a critical role in triggering rejections—can be spotted only by mixing donor and recipient cells and observing their reactions, in a test that takes up to a week. It is no wonder, then, that transplant teams work with the living members of a single family whenever that possibility exists.

To make transplants widely practical, however, researchers had to expand the circle of subjects to include individuals less closely related, others who were not related at all, and still others who were not only unrelated but no longer alive. The researchers did exactly that, not so much by fooling the immune system as by suppressing it. Irradiation of the body, generally with X-rays, has that effect—it kills antibodies and the organs that produce them. Drugs such as prednisone and azathioprine do it, too, and the development of such drugs is one of the growing points in transplant research.

Other research proceeds along more radical lines. Scientists at the University of California at San Francisco, the world's foremost transplant center, found an extraordinary phenomenon: Blood transfusions from the kidney donor to the patient, given before the operation, somehow desensitized the patient's immune system and vastly increased the chances that the transplanted kidney would survive.

The ability to control the immune system and thus prevent rejection of kidneys donated by unrelated subjects has greatly expanded the supply of kidneys available for transplant, although organs from living blood relatives still have the best chance of survival. But suppressing the immune system is dangerous. In such a depressed state the system does not fight off transplants, but neither can it work well against its true enemies—bacteria, viruses and fungi. As scientists learn to control the immune system, the greatest danger after a transplant is coming to be not rejection but infection.

With kidney transplants almost routine, surgeons are turning to parts of the digestive tract hitherto inaccessible to transplant. Some spleen, pancreas and liver transplants have been done. Farther in the future are artificial digestive organs implanted in the body—indestructible, efficient devices, free of the need for donors and unaffected by the attacks of the immune system. An artificial pancreas may be the first such organ to be practicable. Kidneys are another candidate; as artificial kidneys are made smaller and more portable, scientists approach the goal of installing them in the body. These steps point the way to the time when most of the grave defects in the digestive system can be repaired, and the system can function normally in everyone. ✳

A gift of life for an ailing son

Seven-year-old John McMaster III had his heart set on a springtime of Little League baseball. Unfortunately, his kidneys would not cooperate. The Maryland youngster came home from an Easter party on Good Friday, complaining of a stomach-ache. His mother thought she knew the problem: ''Too many candies,'' she told him.

''The next day,'' she recalled, ''he went and played baseball. Then Easter Sunday he felt bad again. His stomach still hurt, and now so did his head.'' She took him Tuesday to a pediatrician; by then John's thighs were covered with purple blotches and his stomach was distended. The doctor had him rushed to Johns Hopkins Hospital in Baltimore.

Blood tests taken there showed trouble: John had a rare blood disease, hemolytic uremic syndrome. In it, thin protein threads form unusual deposits—''like wire mesh,'' said one doctor—in blood vessels, especially in the kidneys. As red blood cells pass through the meshwork, they are sheared and badly damaged; in about 5 per cent of the cases, the kidneys, clogged by the deposits and starved for normal blood, stop working altogether, so that wastes and fluid accumulate. John McMaster was one of that unlucky 5 per cent.

At first John's doctors hoped that the disease would reverse itself and that he would return to normal. Meantime, they severely limited his intake of fluids, salt, protein—the substance clogging his blood vessels—and chemicals such as calcium and potassium. Even with the special diet, however, waste products built up in his blood. To cleanse his system, he was put on peritoneal dialysis *(right)*.

This treatment is an artificial method of doing what the kidneys do naturally. It takes advantage of the fact that when the kidneys fail, the abdominal cavity, enclosed by the saclike membrane called the peritoneum, becomes a repository for abnormal amounts of wastes and fluids. They can, in effect, be rinsed out. A saline solution is injected into the abdominal cavity through a tube penetrating the peritoneum. The rinsing fluid dilutes the waste fluid, then the mixture is drawn off. The procedure is repeated until the extracted fluid is virtually free of toxins.

While diet and dialysis kept John alive, his doctors waited, hoping that his kidneys would repair themselves. They did not: Deprived of nourishment, they had all but died. Yet the disease that had killed John's kidneys was seemingly on the wane. So his doctors mapped their next strategy: to take a kidney from someone else, preferably a relative, and splice it into John's bloodstream. A healthy new kidney would cleanse his system and, if all went well, would be able to resist any remnants of the kidney-killing disease.

PHOTOGRAPHS BY RICHARD ANDERSON

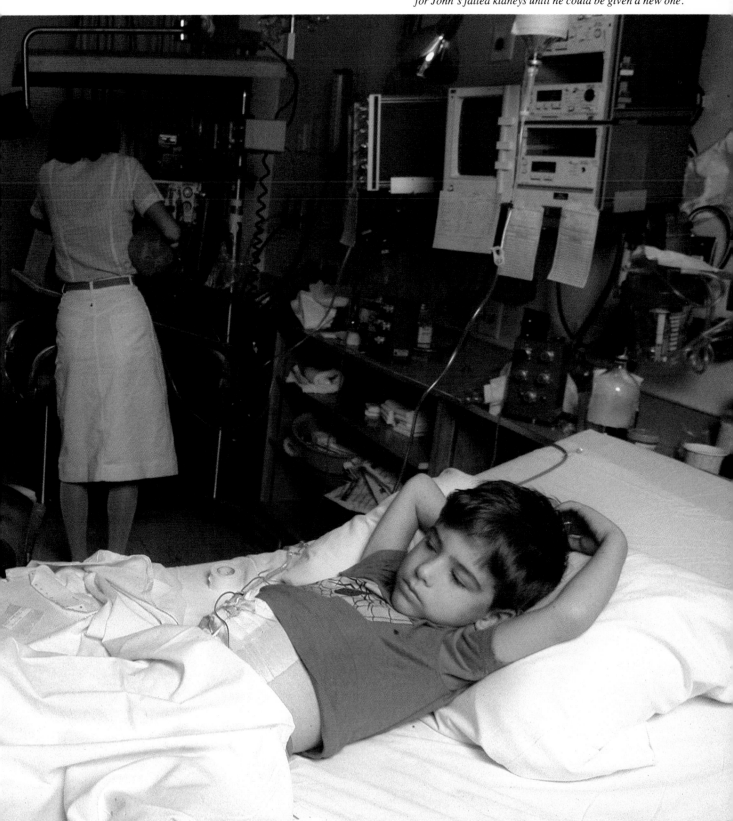

While his mother sits alongside to comfort young John, his system is cleaned by liquid flowing through a catheter inserted into his abdomen. This peritoneal dialysis was performed twice a week in an intensive care unit, usually for 24 hours at a time, to substitute for John's failed kidneys until he could be given a new one.

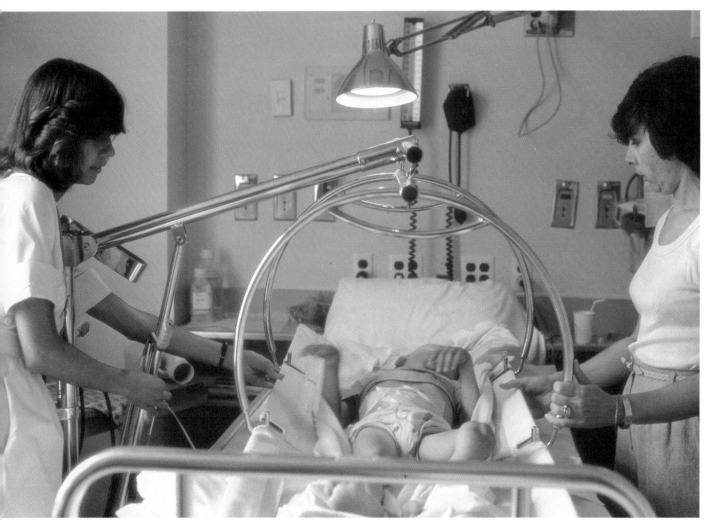

*Suspended in the canvas sling of a special scale, John
is weighed immediately after dialysis. The measurement was
compared with one taken beforehand, enabling doctors to
monitor the amount of waste drawn off. On this day four pounds
of waste fluid were removed from the 40-pound boy.*

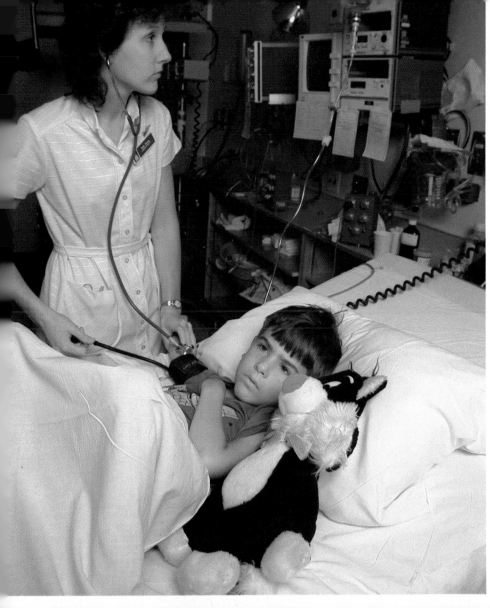

His stuffed cat Sylvester beside him, John has his blood pressure measured. High blood pressure is a frequent side effect of hemolytic uremic syndrome: The kidneys interpret the reduced blood flow as a sign of low blood pressure; they respond by secreting the enzyme renin, which works in the body to constrict blood vessels and raise pressure. John's pressure had to be regulated by drugs.

A blood sample is drawn from John's arm for comparison with blood from his parents. This helped doctors determine whether mother or father would be the more suitable donor of a replacement kidney (overleaf).

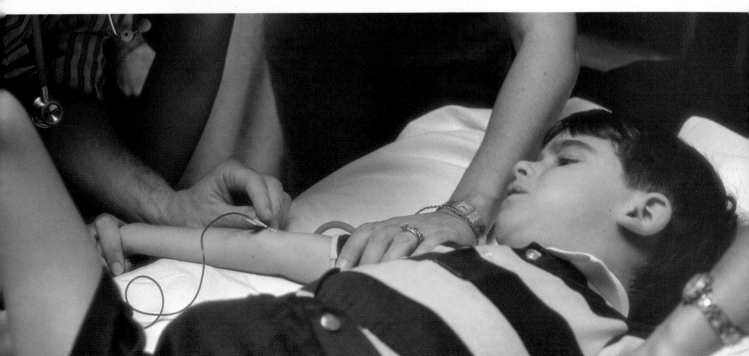

A donor is chosen, a date is set

Both of John McMaster's parents were willing to sacrifice a kidney in order to save their son. Either of them could have withstood the operation and its aftermath: Both were in reasonably good health and had two presumably healthy kidneys—only one is needed to survive. Weeks of blood tests and other examinations suggested that John's 35-year-old father would be a marginally more compatible donor, that his kidney would be less likely to provoke a hostile response from the body defenses of young John's immune system. Such a reaction, known as rejection, would be disastrous: The transplanted kidney, mistaken for a foreign invader, would be killed by the natural processes that are meant to combat disease.

More tests were in store for John's father, principally to confirm the health of his left kidney—it is normally the larger of the two and is the one generally transplanted. If those tests, including the examinations pictured below and at right, had proved disappointing, John's mother would have been called upon. As it turned out, she was not needed. Surgery was set for 8 o'clock on the morning of September 23, almost six months after John McMaster came down with hemolytic uremic syndrome.

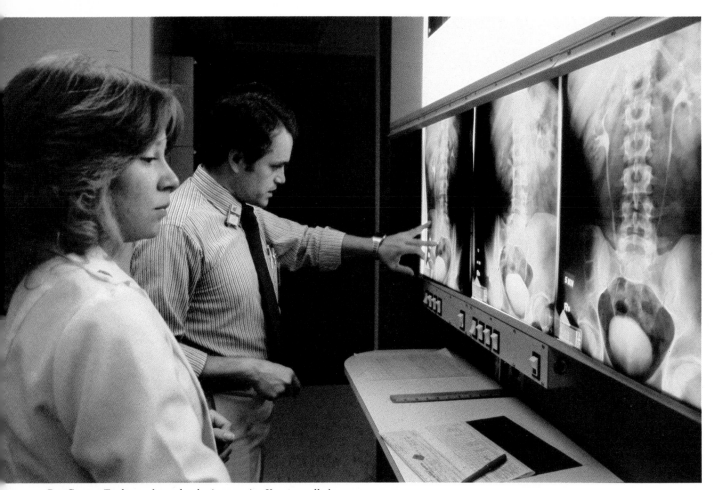

Dr. George Taylor and a technologist examine X-rays, called intravenous pyelograms, of the elder McMaster's urinary tract. To get the pictures, doctors injected special dye into his bloodstream; the dye blocks X-rays as bones do. The pyelogram at far right reveals the funnel-shaped urine-collection points in the kidneys (top) and the long, thin ureters to the bladder (bottom center).

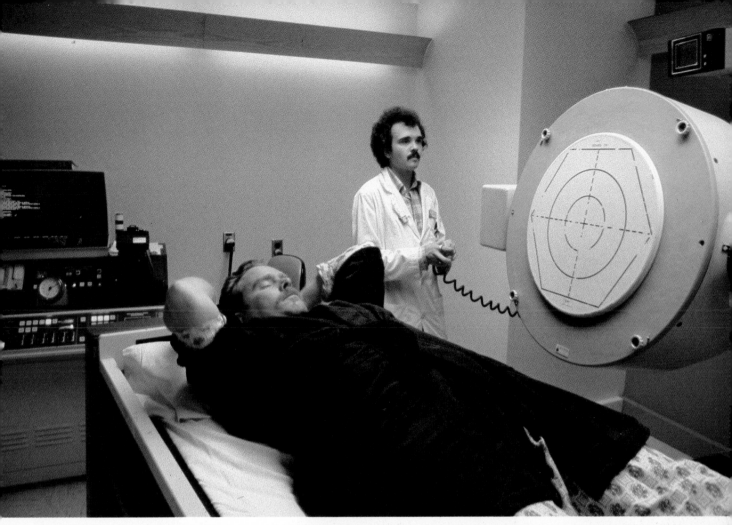

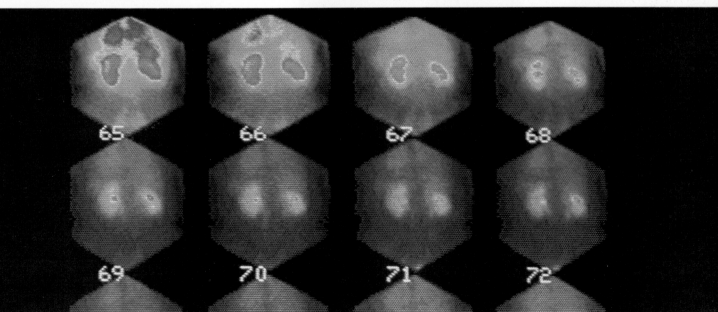

At top, John's father stretches out for a nuclear renal scan while a technologist, using remote control, lowers the huge detector under the patient's back. The scan yields a minute-by-minute view of the kidneys at work: Radioactive material is injected into the blood though an arm vein, the detector picks up the radiation and a computer then converts the emissions into colored images (above).

A series of nuclear renal scans reveals the swift and normal passage of the radioactive tracer through the heart and kidneys of John's father. The colors indicate the concentration of fluid— red is highest, blue lowest. In frame 65, the fluid is concentrated in the heart (top) and the kidneys immediately below; by frame 67, it lies almost entirely in the kidneys.

To surgery, together

The bond between father and son was about to be strengthened in a remarkable way: Part of the elder McMaster's body was soon to become part of his child's. On tables in adjoining operating rooms, hardly 30 feet apart, the two lay sedated but awake. Then, almost simultaneously, intravenous lines dripped anesthetics into their arm veins and they were asleep.

One surgeon worked on John's father, another on the boy. If all went well, father and son would awaken in about four hours—in different rooms, but together as they had never been before.

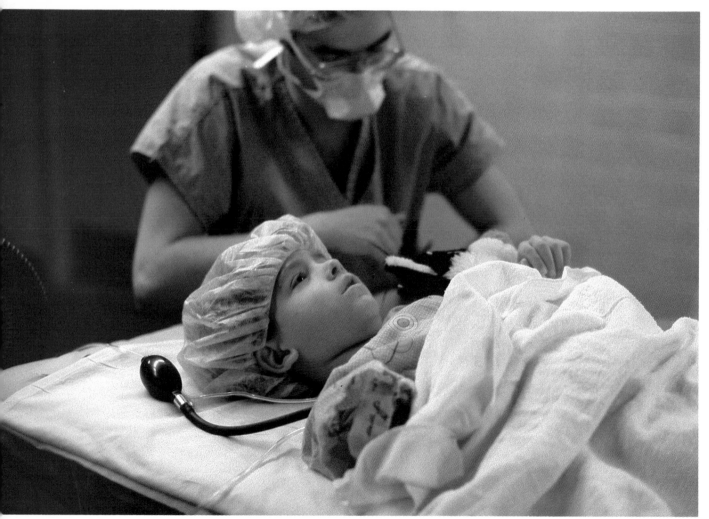

Moments before surgery, an anesthesiologist makes last-minute preparations, working around Sylvester the Cat. Once John was asleep, he was relieved of his mascot. The rubber bulb beside his head is attached to the kind of blood pressure cuff that is used routinely in doctors' offices.

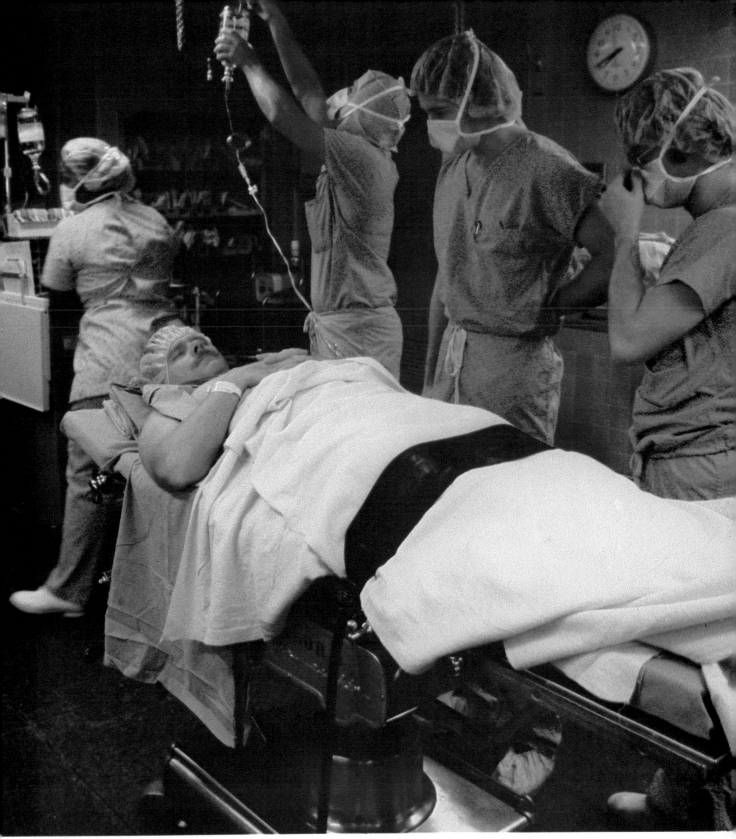

With his eyes shut and his arms folded across his chest, John's father waits on the operating table while an anesthesiologist hitches a bottle of intravenous fluid to an overhead chain. Delivered through a tube and needle in his arm, the fluid supplied water, minerals and drugs during the operation.

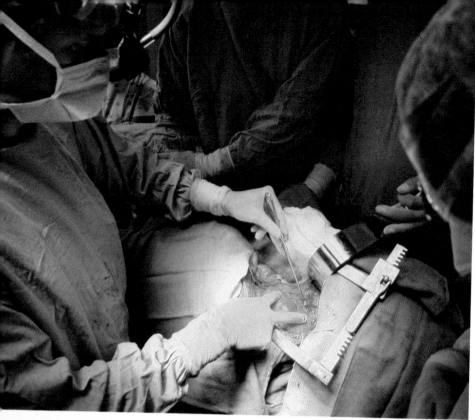

Dr. Fray Marshall, the elder McMaster's surgeon, places two fingers on the firm red kidney (left), just exposed by a 24-inch incision from the hip toward the shoulder blade. With a pointed instrument, he locates a blood vessel that he must cut and stitch off before removing the organ. Below, Dr. James Burdick, young John's surgeon, holds the extracted organ in his left hand. A tube hooked to the kidney's artery and dangling from Dr. Burdick's right hand pumps chilled saline solution through the organ, preserving it and preventing blood clots from forming.

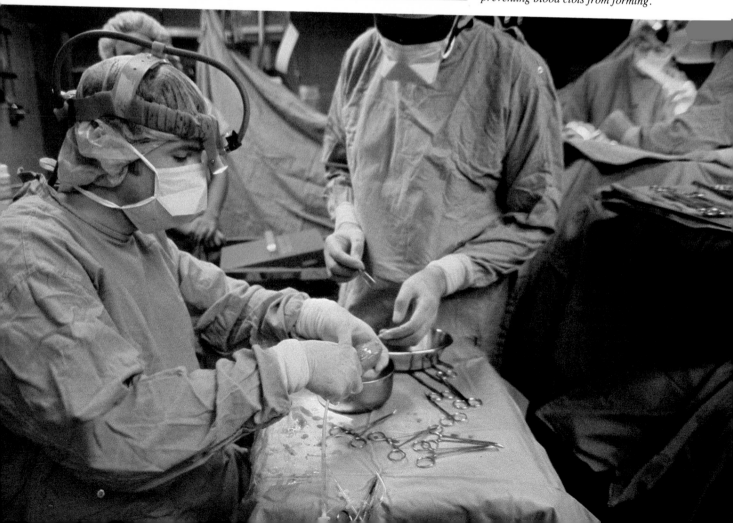

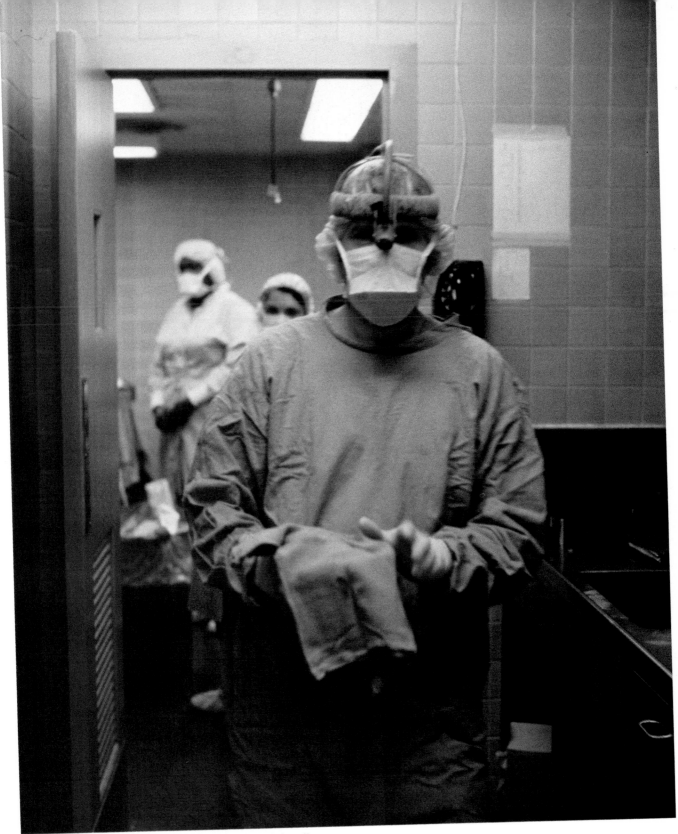

*Carrying the kidney in a towel-covered dish, Dr. Burdick walks
swiftly through the short hallway linking the two operating rooms.
The kidney is bathed in an iced saline solution. Preserved in
this way, it could survive several hours without serious damage.*

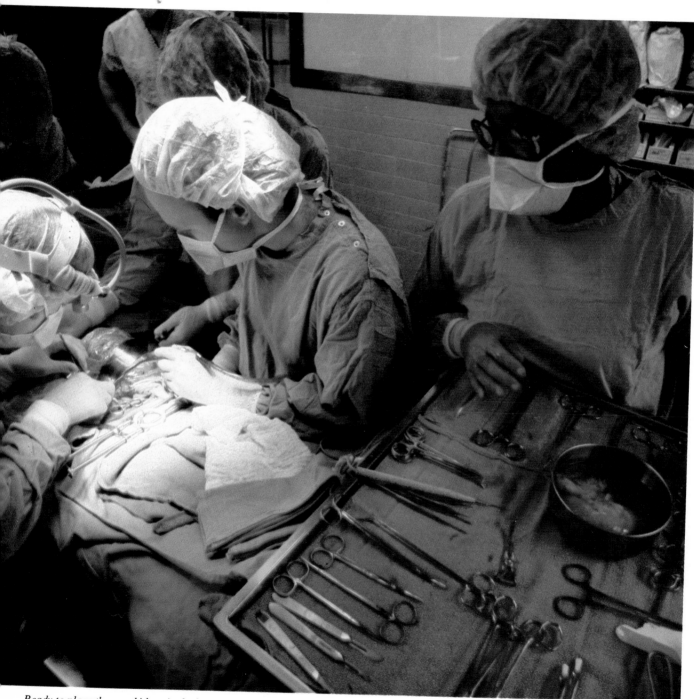

Ready to place the new kidney in the lower part of young John's abdomen, Dr. Burdick makes a final incision in the boy's abdominal aorta, the artery that will be attached to the transplanted organ's own artery. (John's diseased kidneys were left in place, but eventually might be removed.) His new kidney now rests in the dish on the instrument table.

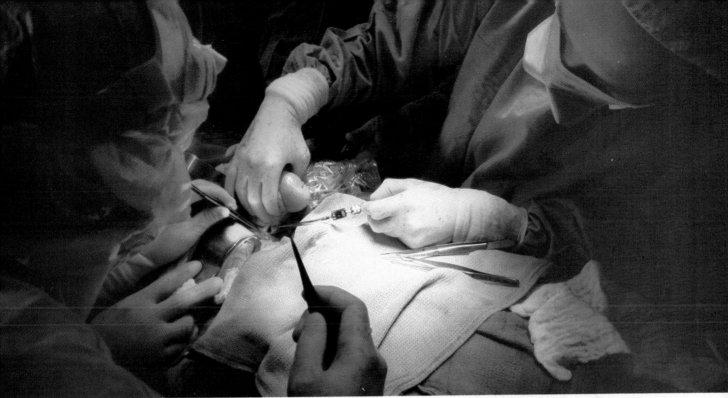

Above, Dr. Burdick sews the kidney's artery to the aorta while an assisting surgeon holds the organ out of the way. Below, an unmistakable sign of success appears: The kidney flushes red with blood after its artery and vein have been connected. The organ produced urine almost at once.

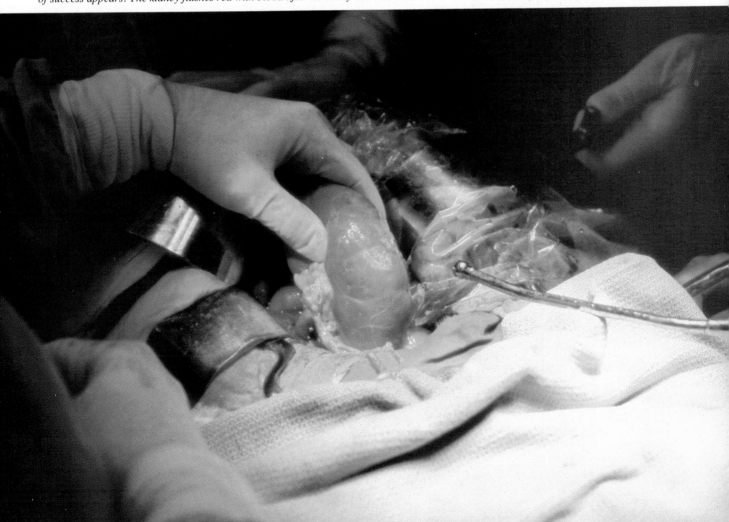

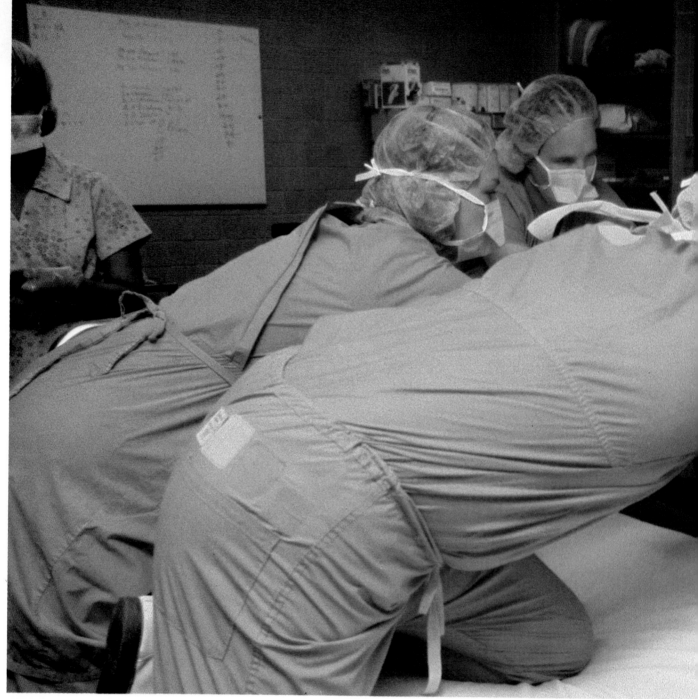

Sylvester the Cat rests on young John's chest—so it will be next to him when he wakes up—as the surgical team gingerly moves John from the operating table to a stretcher. From the operating room, he was wheeled to the pediatric intensive care unit, where he began his recovery.

In intensive care, an anesthesiologist cradles John's head, trying to awaken him. The boy is hooked to standard postsurgical life-support equipment: A tube inserted in John's nose snakes down to his stomach, sucking out digestive juices, and a large, football-shaped bulb, called an ambu bag, is attached to his mouth. If he had had trouble breathing, the doctor would have squeezed it, forcing air into the lungs.

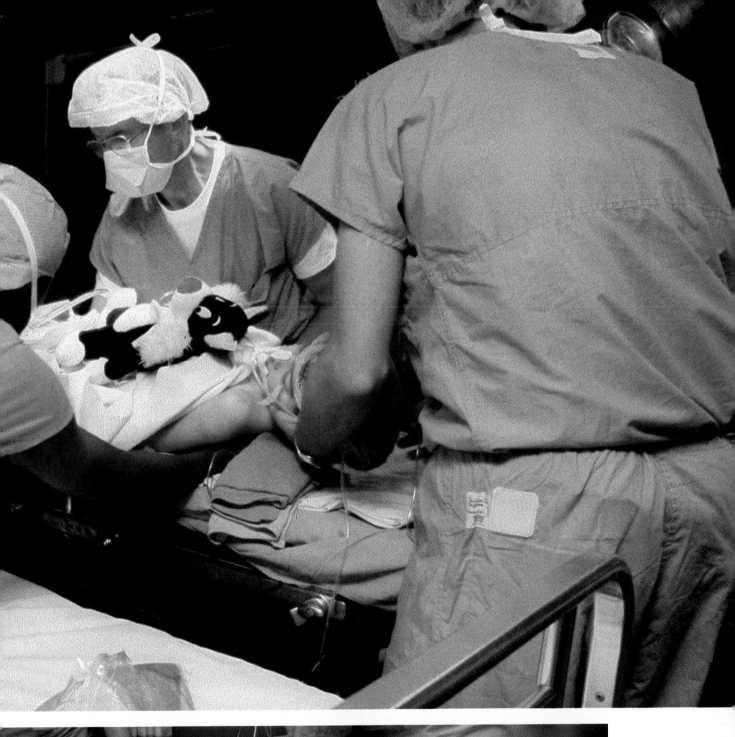
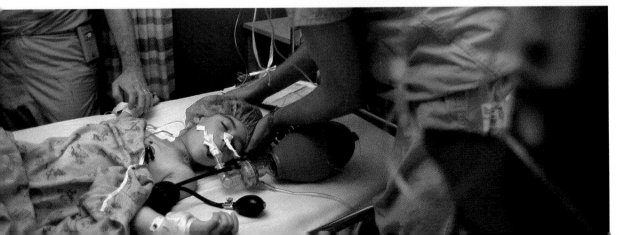

Hazards on the road to recovery

"It was like having a baby, a rebirth. I felt like I had gotten my son back," recalled John's mother of the moment Dr. Burdick told her the operation had been a complete success. But John was not out of danger.

His immune system could turn against the new organ at any time—especially in the first few days. To minimize that chance, he was given azathioprine and prednisone, two drugs that restrain body defenses. The dosages were gradually decreased but would never be eliminated. And while repressing action against the foreign kidney, the drugs also lowered defenses against other aliens, making John vulnerable to illness. Thus, he was kept in temporary isolation, while every action of his transplanted kidney was studied.

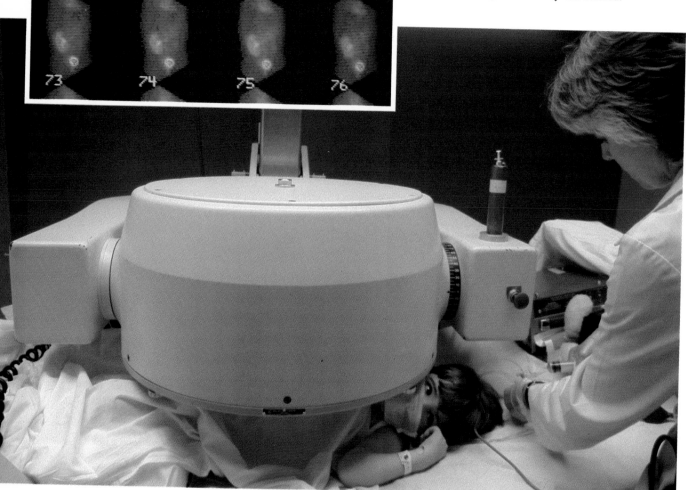

On the day after surgery, John peeks from beneath the detector of a nuclear scan machine. The test, almost identical to that given his father before surgery (page 151), yielded scans (inset) indicating that the new kidney was working well. In frame 65, the tracer (red) is concentrated in John's heart and kidney; by frame 76, most of it has passed to his bladder.

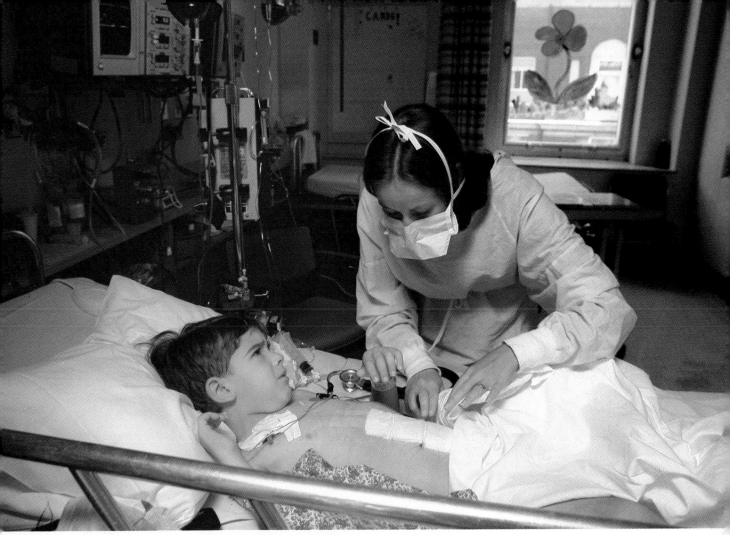

Dr. Barbara Fivush, the pediatric nephrologist who had primary responsiblity for treating John, checks his incision site the day after the operation. John is wired to the monitor above his head, which flashes his heart and breathing rates.

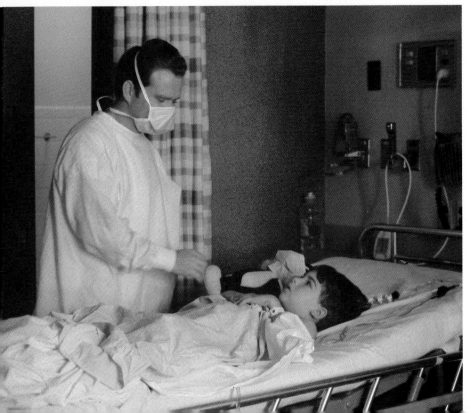

After walking from his own hospital room, pushing a wheeled pole with his intravenous-fluid bottles, the elder McMaster sees his son for the first time after surgery. ''I couldn't have given my kidney to a nicer person,'' he said. He wears a mask to prevent the spread of germs and, in a bit of whimsy, so does Sylvester.

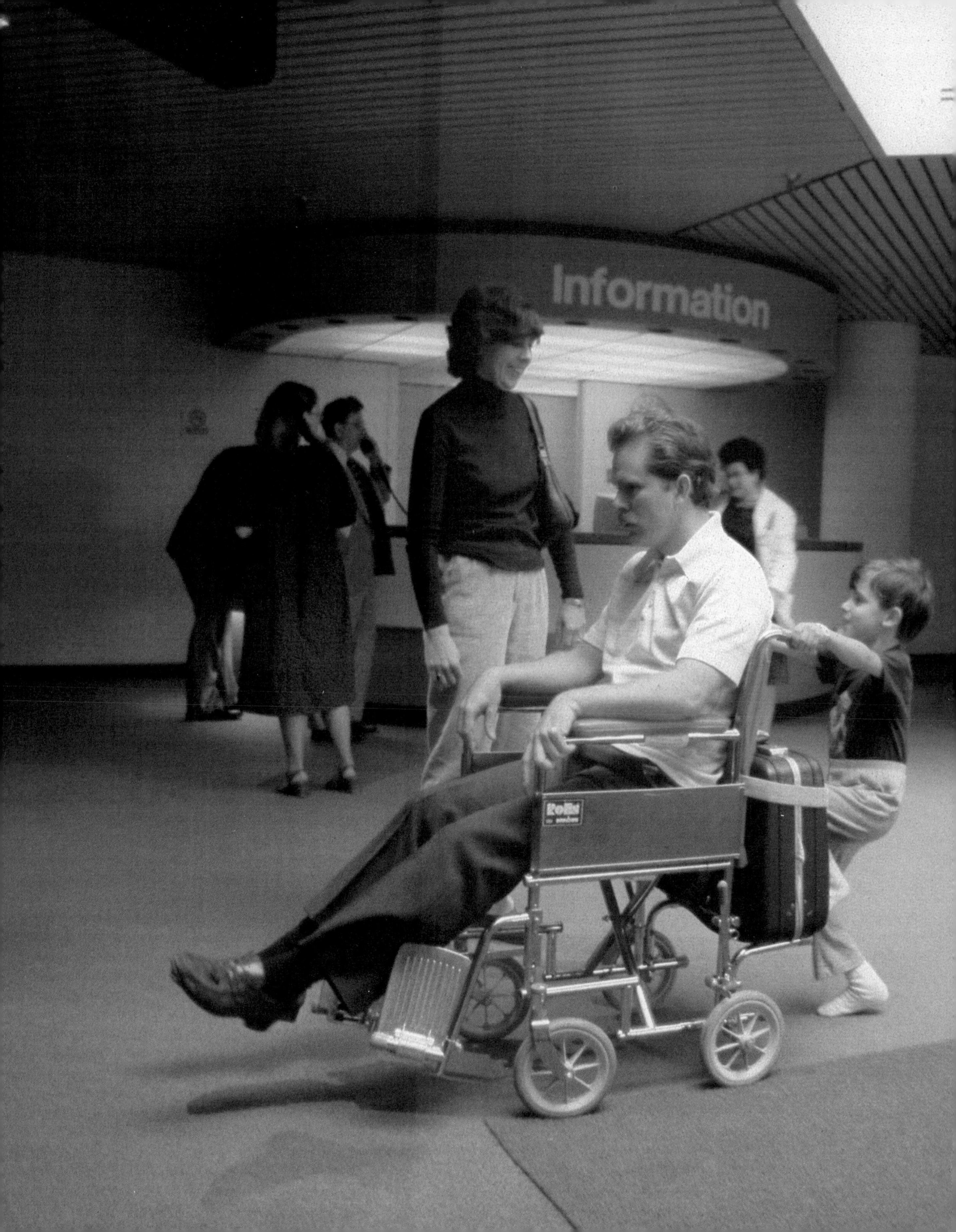
Information

Getting back to (almost) normal

Father and son bounced back fast. Within a week of surgery, Mr. McMaster could go home *(left)*. For little John, the hospital stay was longer—an additional four weeks, principally so that doctors could watch for signs of rejection. He came home from the hospital at the end of October—and went trick-or-treating on Halloween. The following Monday he was back in school. Said his parents: "He's like our old John again."

Successful as his recovery had been, John was not cured. Indeed, he would never be. "I'd never tell anybody it's a pot of gold at the end of the rainbow," said Dr. Fivush of kidney transplants. "There are a lot of problems. The kids always need to be on medication. There's always the chance of rejection. But for someone who really doesn't have kidneys that work, I think it's just miraculous." Of John McMaster's prospects, she said, "We are very optimistic."

Testing his strength, young John tries playfully—and vainly—to tip the wheelchair backward while his father prepares to leave the hospital. (A wheelchair departure is standard for discharged patients.) Above, mother and son take equal pleasure in his first enjoyable meal in many months—the hospital's fried shrimp.

An encyclopedia of symptoms

Usually the body's digestive organs work together in a smoothly orchestrated process that passes food without incident from the mouth to the stomach and then through the intestines. When something goes wrong, however, the symptoms are generally unmistakable; they may include abdominal pain, nausea, vomiting, diarrhea, bleeding, or a change in bowel habits.

The best guides to the origin of the problem are the location or intensity of the pain or discomfort, its characteristic "feel," and the duration of the symptoms. As a rule of thumb, any digestive symptoms that seem unduly severe should send you immediately to the doctor, as should any symptoms that persist for more than a week or that recur frequently. Generally, an episode of bleeding from the rectum should be reported to a physician within 24 hours.

The most common digestive symptoms are listed alphabetically below. The condition that causes each symptom or group of symptoms appears in small capital letters.

ABDOMINAL PAIN. Stomach-aches usually are mild and temporary reactions to dietary indiscretions. However, stomach-aches that are severe, prolonged, recurrent or accompanied by other, more alarming symptoms can signal a serious disorder.

● **Abdominal pain that occurs in the center of the belly,** between the navel and ribs, usually suggests INDIGESTION, a temporary discomfort that is usually related to overeating or eating overly rich foods. Other symptoms include belching or a feeling of fullness. If the pain is accompanied by a burning sensation, either in the pit of the stomach or behind the breastbone, HEARTBURN may be indicated. Unless chronic or severe, these common symptoms seldom demand a doctor's care. Try to avoid the foods that seem to provoke discomfort, eat smaller meals and take antacids.

● **Abdominal pain that occurs suddenly and is accompanied by nausea, vomiting and appetite loss** can have a variety of causes.

If sudden, steady abdominal pain occurs in the pit of the stomach, between the navel and ribs, and is associated with nausea, vomiting and appetite loss, the problem may be GASTRITIS, an inflamed stomach lining. This could be the result of hangover, aspirin sensitivity, infection, or even bacterial contamination of food. For mild symptoms, rest and drink clear fluids. Consult a physician at once if pain is severe or if vomit is tinged with blood.

If sudden, crampy abdominal pain occurs with fever, diarrhea, nausea, vomiting, appetite loss and headache, the cause may be GASTROENTERITIS, an intestinal inflammation that usually lasts for a day or two. Rest and drink extra fluids. See a doctor if symptoms are severe or persistent, or if blood tinges the vomit or stool.

If severe pain in the upper right side of the abdomen is associated with nausea and fever, it may indicate ACUTE CHOLECYSTITIS, an inflamed gall bladder. Consult a physician within a few hours.

If pain in the lower left side of the abdomen is associated with fever and constipation of recent onset, it may suggest DIVERTICULITIS, an infection of pouches that sometimes form in the large intestine. Consult a physician within a few hours.

If pain begins suddenly in the middle of the abdomen, then shifts to the lower right side, the cause may be APPENDICITIS, an infected appendix. Other symptoms include appetite loss, nausea, vomiting, fever and constipation. Consult a physician if steady abdominal pain persists for more than four hours.

If sudden, severe, persistent midabdominal pain occurs with nausea, vomiting and abdominal swelling, it may indicate INTESTINAL OBSTRUCTION, blockage of the bowel, often caused by scar tissue that forms after surgery, injury or infection. Consult a physician immediately.

If severe, steady, midabdominal pain feels as if it is boring through toward the back and is associated with nausea, vomiting and abdominal swelling, the problem may be ACUTE PANCREATITIS, an inflamed pancreas. Consult a physician immediately.

If severe pain in the upper right side of the abdomen occurs with fever, chills and jaundice (yellow skin), the cause may be ACUTE CHOLANGITIS, obstructed and inflamed bile ducts. Consult a physician immediately.

If sharp, severe abdominal pain occurs suddenly and is felt on one side, then spreads into the lower abdomen or groin, the cause may be KIDNEY STONES, hardened mineral deposits in one of the tubes that connect the kidneys to the bladder. Consult a physician immediately if pain is intense, within a few hours if it is not.

● **Abdominal pains that are recurrent or chronic** afflict many women during menstruation or at the time an egg is released by an ovary, but such pains can also indicate various digestive ills in men and women.

If recurrent abdominal pain is sharp but intermittent and is felt around the navel, or on the left side of the abdomen or chest, the cause may be INTESTINAL GAS, air stretching the bowel wall. Pain is often relieved by a bowel movement or the passing of gas (see FLATULENCE). Avoid gas-producing foods such as beans, peas, onions and cabbage. See a doctor if pain is severe or persistent.

If recurrent crampy or steady pain occurs in the lower right side and is associated with chronic or intermittent diarrhea, flatulence or weight loss, the cause may be CROHN'S DISEASE, an inflammatory bowel disease. Consult a physician within a day or so.

If recurrent pain in the pit of the stomach, between the navel and ribs, is experienced as a gnawing or burning ache in a very small area that you can cover with one or two fingers, it may suggest an ULCER, a sore in the lining of the stomach or upper intestine. Pain usually occurs within several hours of a meal, when the stomach is empty, and will often be relieved—temporarily—by eating or taking antacids. Consult a physician within several days.

If recurrent midabdominal pain in an elderly individual regularly follows eating, the problem may be INTESTINAL ANGINA, poor blood flow to the bowel. Consult a physician within a day or so.

• **Persistent pain that occurs in the upper left abdomen** and is associated with chronic constipation, or with alternating episodes of constipation and diarrhea, may indicate IRRITABLE BOWEL SYNDROME, irregular intestinal function that is often related to psychological stress. This is also called SPASTIC COLITIS or MUCOUS COLITIS, and stools may be unusually thin or contain mucus. Consult a physician within a few days.

• **Pain in the upper right side of the abdomen** that radiates to the back or to the right shoulder may suggest BILIARY COLIC, hardened mineral deposits in the bile duct. Pain begins suddenly, persists for several hours and passes. Nausea is often present. If symptoms disappear, you can wait until the next day to see the doctor, but if they persist or worsen, consult a physician immediately.

ABDOMINAL SWELLING. True abdominal swelling is different from a feeling of fullness or a sensation of bloating that may accompany INDIGESTION, the commonplace discomfort related to eating. Swelling is comparatively rare and usually develops gradually, over several weeks. It can be difficult to observe, particularly in a person who is overweight.

• **Painless abdominal swelling** accompanies several serious ills.

If abdominal swelling causes no pain but is accompanied by a sensation of pulling or stretching in the flank or groin, the cause may be ASCITES, excess fluid in the abdominal cavity. The condition is frequently caused by a liver disorder such as CIRRHOSIS, the severe scarring that usually is due to alcoholism. Such liver disorders may also produce the yellowed skin of jaundice. Consult a physician within a few days.

If abdominal swelling causes no pain but is associated with shortness of breath, the problem may be CONGESTIVE HEART FAILURE, inadequate heart output; or UREMIA, a chronic kidney disease that also can cause appetite loss, nausea or behavioral changes such as sleep and memory disturbances. Swelling of the legs may also be present. Consult a physician immediately.

• **Abdominal swelling that occurs suddenly and is accompanied by severe abdominal pain** can suggest one of two distinct disorders. Both are serious.

If abdominal swelling is associated with steady pain in the middle of the abdomen that bores through toward the back, as well as with nausea, vomiting and sweating, the problem may be ACUTE

PANCREATITIS, an inflamed pancreas. This condition is often due either to GALLSTONES, hard mineral deposits in the gall bladder, or to alcohol abuse. Consult a physician immediately.

If abdominal swelling is associated with crampy pain in the middle of the abdomen and with vomiting, it may suggest INTESTINAL OBSTRUCTION, blockage of the bowel, often caused by scar tissue that forms after surgery, injury or infection. Vomit may be green or orange-brown and have a foul odor. Consult a physician immediately.

ANAL DISCOMFORT. Pain around the anus, painful defecation, itching, or bleeding from the anus are symptoms of many so-called PERIANAL DISEASES, disorders of the sensitive anal tissues. Rarely is a doctor's attention required, and then only if the symptoms are severe or persistent, or if blood is noted on or in the stool.

• **Anal discomfort or irritation that may be accompanied by itching or burning, by pain on defecation or by bleeding** usually indicates HEMORRHOIDS, swollen rectal or anal veins. (A firm, painful anal lump normally suggests a THROMBOSED HEMORRHOID, a blood clot in a swollen vein.) Symptoms may be aggravated by straining at stool, by constipation, by excess weight or by pregnancy. Hot baths can help reduce pain and swelling. Keep the affected area clean and dry. Soften stools by drinking extra liquids and eating foods that contain fiber, such as fresh fruits, vegetables and bran. Hemorrhoid ointments, creams, pastes and gels can help relieve discomfort, principally by coating the affected area and protecting it from irritation. Occasional bleeding is seldom serious, particularly if blood is bright red in color (indicating an origin very close to the anus); however, any rectal bleeding should be called to the attention of a doctor, for it may signal a serious ailment, including cancer.

• **Painful defecation that is accompanied by blood or pus** may indicate an ANAL FISSURE, irritated or eroded anal tissues; an ANAL ABSCESS, a bacterial infection of anal tissues; or an ANAL FISTULA, a tubelike passage that forms between the intestine and the anal skin. These conditions may in turn be indications of such serious disorders as CROHN'S DISEASE or CHRONIC ULCERATIVE COLITIS, two inflammatory bowel diseases that can also cause bloody diarrhea, weight loss and fever. Consult a physician within a day or so.

• **Multiple small growths on the anal skin** may indicate CONDYLOMA ACCUMINATA, anal warts caused by a viral infection. Consult a physician within a few days.

• **Severe, persistent anal itching in an adult** may suggest PRURITIS ANI, a psychosomatic problem that most often strikes middle-aged men. Itching may be nearly constant, worsening at night and during hot weather. Consult a doctor within a few days.

If persistent anal itching occurs in a child, the cause may be ENTEROBIASIS, pin-worm infestation of the bowel and anus. Itching is worse at night and can lead to insomnia, restlessness and bed-wetting. Consult a physician within a day.

APPETITE LOSS. Loss of the desire to eat, called anorexia, is most often short-lived and related to minor emotional upsets or to changes in daily routine. Temporary appetite loss often accompanies minor illnesses such as INFLUENZA, a viral illness; or GASTROENTERITIS, an inflammation of the intestines. Such illnesses usually pass within several days. Appetite loss that is persistent or causes weight loss is another matter and demands medical attention.

• **Chronic appetite loss that is accompanied by weight loss** may indicate CANCER or UREMIA, a chronic kidney disease that may also cause nausea and behavior changes such as sleep or memory disturbances. Consult a physician.

• **Chronic appetite loss that occurs in an adolescent, particularly a girl,** may suggest ANOREXIA NERVOSA, a psychological disorder in which severe dieting results in prolonged, often dangerous weight loss. Often the person does not perceive the life-threatening behavior as a problem. Alternating episodes of fasting and binge-eating can also occur. Consult a physician.

• **Appetite loss that occurs suddenly and is accompanied by steady or severe abdominal pain, nausea, vomiting or fever** may suggest one of several acute abdominal emergencies. The most common of these is APPENDICITIS, an infection of the appendix. Pain may begin in the middle of the abdomen, then intensify and move toward the lower right side. Consult a physician immediately if such symptoms persist or if abdominal pain worsens over four to six hours.

• **Appetite loss that is accompanied by fever, headache, fatigue, sore throat, cough, aching muscles and joints, and the yellow skin of jaundice** suggests HEPATITIS, inflammation of the liver caused by a virus or by chemical poisoning, often from alcohol. The appetite loss is frequently accompanied as well by nausea, vomiting and changes in sense of smell or taste. Consult a physician immediately.

BACK PAIN. Back pain is most often caused by disorders of the frame such as poor posture, muscle strain or a SLIPPED DISC, a tear of the spongy material that separates the spinal bones. In these conditions, pain is worsened by movement and relieved by rest. Back pain can also be a symptom of disease in the digestive tract or kidneys. If this is the case, pain is usually unrelated to movement or rest and often feels as though it is moving through the belly and into the back. Consult a physician when experiencing any chronic back pain or a severe back pain of recent onset.

• **Back pain that occurs suddenly and is accompanied by ab-**dominal pain, abdominal swelling, nausea, vomiting, sweating or incapacitating weakness** may indicate ACUTE PANCREATITIS, an inflamed pancreas; or a PERFORATED ULCER, a hole in the stomach or upper intestine. The pain may feel as though it is boring through the abdomen toward the back. Consult a physician immediately.

• **Back pain that occurs suddenly in the right shoulder or shoulder blade** and that is accompanied by nausea and a steady ache in the upper right side of the abdomen may indicate BILIARY COLIC, hardened mineral deposits in the bile duct. Pain and nausea usually begin about an hour after eating, persist for several hours and then subside. If the pain passes, you can wait until the next day to see the doctor; consult a physician immediately, however, if it gets worse, persists for more than four hours or occurs with jaundice (yellow skin), fever or chills.

• **Back pain that occurs suddenly and is accompanied by fever, chills, nausea, vomiting and diarrhea** may suggest ACUTE PYELO-NEPHRITIS, a bacterial kidney infection. Consult a doctor at once.

• **Chronic back pain that is dull, aching and unrelated to movement** can be an indication of PANCREATIC CANCER or of an AORTIC ANEURYSM, a bulge in an abdominal artery that can press against the spine. Consult a physician within a few days.

BELCHING. Belching is usually brought on because air has been swallowed. This may occur while eating or drinking. Belching can also result from nervousness or anxiety. Belching is rarely a symptom of serious digestive disease and, in fact, it sometimes helps relieve INDIGESTION, the discomfort of after-eating fullness.

• **Belching that is accompanied by a warm or burning feeling behind the breastbone** may indicate HEARTBURN, distention or irritation of the tube that carries food from the mouth to the stomach. Some people find that belching relieves this pain, while in others the pain occurs on belching. HEARTBURN is seldom a serious problem. Yet because it can have several underlying causes—including HIATAL HERNIA, a protrusion of the stomach through the diaphragm—consult a physician if it is a daily problem. If symptoms are mild or occasional, treat the condition with antacids, eat smaller portions at meals and try to lose weight—reduced weight lessens pressure on the stomach.

BLEEDING. *See ANAL DISCOMFORT, DIARRHEA, REGURGITATION, STOOL CHANGES, VOMITING BLOOD*
BOWEL-HABIT CHANGES. *See ANAL DISCOMFORT, CONSTIPATION, DIARRHEA, STOOL CHANGES*

CHEST PAIN. Chest pain is a common but disconcerting complaint that is more often caused by a digestive disorder or a muscle strain than by heart disease. It often indicates simple intestinal gas. Such pain may be either sharp and stabbing or crampy and intermittent. Relief may occur with the passage of gas or on defecation.

• **Severe chest pain that comes on suddenly during exertion** and that forces you to stop all activity may indicate ANGINA PECTORIS, a brief episode of insufficient blood flow to the heart; or a MYOCARDIAL INFARCTION (HEART ATTACK), a complete blockage of part of the heart's blood supply. The pain of a heart attack may be unremitting and is often described as a squeezing, gripping or crushing heaviness or tightness. Nausea, sweating and shortness of breath may occur. Angina demands medical attention but it is not an emergency; a heart attack is life-threatening—call an ambulance if it is suspected.

• **Burning chest pain that is localized behind or beneath the breastbone** may arise in the heart from angina or infarction *(above),* but it normally originates in the esophagus—the tube that carries food from mouth to stomach.

If burning pain first occurs in the middle of the chest and then spreads upward in waves, the cause may be HEARTBURN, a painful distention or irritation of the lower part of the esophagus. Some people have learned that this pain will invariably result after they eat certain foods. For others it begins unexpectedly and may be induced by alcohol or aspirin. Belching or swallowing sometimes relieves this pain by helping to relax the lower part of the esophagus. Take antacids, eat light meals and lose weight—weight loss reduces pressure on the stomach. Consult a physician if pain is severe or recurrent.

If burning chest pain recurs after heavy meals and if it is worsened by bending over, lying down or straining, it may suggest HIATAL HERNIA, a protrusion of the stomach through a hole in the diaphragm; or REFLUX ESOPHAGITIS, irritation of the lining of the esophagus by stomach juices. Regurgitation of food into the mouth—and coughing if such food slips back into the windpipe—may accompany either of these conditions. Many people plagued by what they think is simple heartburn actually suffer from these more complex disorders. Consult a physician if such chest pain is a frequent problem. Otherwise, take antacids, eat light meals and lose weight to reduce pressure on the stomach. Elevating the head of your bed six to eight inches may also help.

• **Chest pain that is accompanied by difficulty swallowing, a sensation of food sticking in the chest, or a feeling of fullness behind the breastbone** may suggest one of several disorders of the esophagus, such as ACHALASIA, a dilated and ineffective food tube; DIFFUSE ESOPHAGEAL SPASM, a painful muscular contraction of the food tube; or ESOPHAGEAL CANCER, a tumor in the food tube. Consult a physician immediately if the pain is severe, within a day if it is not.

CHOKING. When food blocks the windpipe, the victim can neither breathe nor speak and is unable to tell you that he is choking. If during a meal a person suddenly becomes speechless and then grabs at his chest or neck as if suffering a heart attack, he may instead be choking (a heart attack victim usually will be able to say that he is having chest pain or shortness of breath). Choking is an emergency—death follows swiftly—and it calls for immediate action by someone trained in first aid. If the so-called Heimlich maneuver *(page 13)* is used to force out the blockage, a physician should be consulted soon afterward to make sure that no damage was done to the lungs or ribs.

CONSTIPATION. Constipation is neither as common nor as serious as many people think. Individuals vary widely in their frequency of bowel movements: Normal function can range from several times a day to once in several days. Temporary changes in bowel habits caused by minor upsets in daily routine do not necessarily indicate constipation and should not be a cause for worry or for the use of laxatives. Normally all that is required is a change in diet: Drink extra liquids and eat foods that contain fiber, such as fresh fruit, vegetables and bran. Both bulk-forming laxatives containing psyllium and stool softeners containing a docusate compound can be effective, but they should be used only occasionally and then only for short periods. Sometimes constipation signals a disorder of the digestive system.

• **Chronic constipation that is a consequence of or that occurs with painful bowel movements** may suggest a PERIANAL DISEASE, a disorder of the anal tissues. Usually the problem is HEMORRHOIDS, swollen veins in the anus (see ANAL DISCOMFORT). Consult a physician if such symptoms are unduly severe or persistent.

• **Chronic constipation accompanied by crampy pain in the left side of the abdomen** may suggest IRRITABLE BOWEL SYNDROME, irregular intestinal function often related to psychological stress. Also called SPASTIC COLITIS or MUCOUS COLITIS, the condition is often associated with alternating episodes of diarrhea and constipation. Consult a physician within a few days.

• **Constipation that occurs suddenly and is accompanied by fever and a steady pain in the lower abdomen** may suggest DIVERTICULITIS, an infection of the pouches that sometimes form in the large intestine. If there is also nausea, vomiting and appetite loss, the problem may be APPENDICITIS, an infection of the appendix. Consult a physician if the pain worsens, or if it persists for more than four hours.

• **Constipation that is accompanied by very dark, unusually thin or blood-tinged stools** may indicate INTESTINAL CANCER. Consult a physician within a day for any persistent change in bowel habits or at any sign of rectal bleeding.

DIARRHEA. This is one of the most familiar signals of trouble in the digestive tract. It may be caused by emotional stress, by excessive eating, drinking or smoking, or by a brief infection of the digestive system. Usually it is nothing to worry about and will vanish within 24 hours. When diarrhea strikes, avoid solid foods—they can aggravate the condition—take a rest and drink extra fluids, for dehydration frequently accompanies the episode. Nonprescription medicines containing polycarbophil or kaolin and pectin can be effective. Consult a physician if diarrhea is a recurrent problem, if it persists for more than two days, if it occurs in a child or if it is accompanied by other, more worrisome symptoms—especially blood, pus or mucus in the stool.

● **Diarrhea that occurs suddenly and with additional symptoms** can have a variety of causes.

If sudden diarrhea is accompanied by abdominal cramps, fever, headache, loss of appetite, vomiting, nausea or fatigue, the cause may be GASTROENTERITIS, an inflammation of the intestines. Varieties of this ailment include SHIGELLOSIS, DYSENTERY, TRAVELER'S DIARRHEA and the misnamed 24-HOUR FLU. If blood or mucus are noted in the stool, consult a physician within several hours; otherwise, follow the standard diarrhea treatment *(above)*.

If sudden diarrhea is accompanied by nausea, vomiting and abdominal cramps, the cause may be FOOD POISONING, from bacterial contamination of food. Symptoms usually last for a day or so and generally affect most or all of the persons who have shared a meal. Consult a physician if symptoms persist. If such symptoms are swiftly followed by difficulty seeing, swallowing or speaking, or by weakness or paralysis, the cause may be BOTULISM, food poisoning due to *Clostridium* bacteria. Get medical attention immediately if this disorder is suspected.

If sudden diarrhea is accompanied by severe pain in the lower abdomen and rectal bleeding, particularly in an older person, the cause may be ISCHEMIC COLITIS, insufficient blood flow to the bowel. Consult a physician immediately.

● **Diarrhea that is present continuously or that recurs frequently** can have a variety of causes.

If chronic diarrhea is associated with fever, abdominal cramps or pain in the lower abdomen, the cause may be either CROHN'S DISEASE or ULCERATIVE COLITIS, two inflammatory bowel diseases that often leave blood, pus or mucus in the stool. Appetite loss, weight loss and fatigue sometimes occur. Consult a physician within a few days if symptoms are mild, but within a few hours if blood, pus or mucus is noted in the stool.

If chronic diarrhea alternates with episodes of constipation and is associated with nausea or crampy pain in the lower abdomen, the cause may be IRRITABLE BOWEL SYNDROME, also called SPASTIC COLITIS or MUCOUS COLITIS. The condition signals irregular intestinal function, and is often related to psychological stress. Symptoms are more frequent in the morning and usually do not occur at night. Consult a physician within a few days.

If chronic diarrhea is accompanied by stools that are unusually bulky, foul-smelling or greasy, or that are pale yellow, the cause may be LACTOSE INTOLERANCE, the inability to digest milk; or GLUTEN INTOLERANCE, the inability to digest proteins in foods made from wheat, rye, barley and oats. Associated symptoms can include abdominal cramps, bloating, flatulence, loss of appetite and weight, and easy bruising. Try to avoid suspected foods, and consult a physician within several days.

If chronic diarrhea is accompanied by chalk-colored stools and frequent coughing fits or respiratory infections in a child, the cause may be CYSTIC FIBROSIS, a genetic disorder of mucous glands. Consult a physician within a day or so.

FEVER. A rise in body temperature above 98.6° F. often accompanies such abdominal symptoms as pain, appetite loss, nausea, vomiting and diarrhea. Bed rest, acetaminophen and cool baths will usually lower temperature. Avoid aspirin if your stomach is upset—it will reduce fever but can aggravate digestive symptoms. Sip cool water to replenish fluids lost through perspiration. Consult a physician if fever exceeds 102° F., if a lower fever persists for more than a week, or if other worrisome symptoms are present. These can include abdominal pain, diarrhea, constipation, nausea, vomiting, headache, appetite loss, fatigue and the yellow skin of jaundice.

FLATULENCE. Although the passing of gas through the rectum may prove an occasional embarrassment, it is seldom a sign of serious digestive disease. Gas in the bowel can come from three sources—swallowed air, air-filled foods such as carbonated beverages, or the decomposition of food. Some people swallow air while eating, others swallow it compulsively because of anxiety. Certain foods are known to encourage bacterial gas formation. These include fatty foods, air-containing concoctions such as milk shakes or meringues, and many vegetables, notably beans and members of the cabbage family. To avoid flatulence, avoid these foods; professional attention is needed only if flatulence is a constant worry, or if it is accompanied by severe pain or other alarming symptoms.

● **Flatulence that is accompanied by crampy abdominal pain and chronic diarrhea,** with stools that are unusually bulky, foul-smelling, greasy or pale may suggest LACTOSE INTOLERANCE, the inability to digest milk; or GLUTEN INTOLERANCE, the inability to digest proteins in wheat, rye and some other grains. Try eliminating the suspected foods from the diet, but you should also consult a physician within several days.

HEAD PAIN. Head pain is often associated with digestive disorders such as GASTROENTERITIS, an inflammation of the intestines; or HEPATITIS, an inflamed liver. In such illnesses there will usually be other symptoms, including nausea, vomiting, diarrhea or the yellow skin of jaundice. Head pain can also be caused by injury to the head and by disorders of the brain; these may produce secondary digestive symptoms, the most common of which are nausea and vomiting. If nausea follows a blow to the head, consult a physician immediately.

If throbbing head pain occurs on one side of the head and is accompanied by nausea, vomiting and sensitivity to light and sound, it may indicate COMMON MIGRAINE, a painful dilation of blood vessels supplying the head. This is sometimes called SICK HEADACHE because of the abdominal distress that so often accompanies the head pain. If the symptoms are preceded by a 15- to 30-minute period during which unusual sights and feelings are perceived, the cause may be CLASSIC MIGRAINE. Consult a physician if symptoms are severe or recur frequently.

JAUNDICE. This yellowing of the skin, which often comes on so gradually it is first noticed by friends, is not a disease in itself but rather an indication that something is wrong with either the liver, gall bladder or pancreas. Jaundice can change the whites of the eyes to yellow and is often accompanied by tea-colored urine, pale stools and generalized itching. Consult a physician at the first sign of jaundice.

● **Jaundice that occurs several days after the abrupt onset of nausea, vomiting, appetite loss, fever and weakness** may indicate HEPATITIS, inflammation of the liver. Additional symptoms include headache, muscle or joint pain, sore throat, cough or changes in the sense of taste or smell.

● **Jaundice that develops gradually and is accompanied by appetite loss, fatigue and abdominal swelling** may suggest CIRRHOSIS, scarring of the liver most often caused by chronic alcoholism. Other symptoms include weight loss, menstrual irregularities in women and decreased testicle size and breast enlargment in men.

● **Jaundice that occurs suddenly and is accompanied by fever, shaking chills and pain in the upper right side of the abdomen** may indicate ACUTE CHOLANGITIS, an obstruction of the bile duct by hardened mineral deposits (gallstones).

NAUSEA AND VOMITING. These symptoms are among the most common signs of digestive disorders. They may be accompanied by appetite loss, pale or clammy skin, or increased saliva production. For brief episodes of nausea or vomiting, rest and sip water or chew on ice chips. Professional attention is needed only if either condition is severe, recurrent or persistent, if it affects a child, or if it is accompanied by other warning symptoms.

● **Nausea and vomiting that occur with crampy abdominal pain, diarrhea, fever, fatigue and headache** may be signs of GASTROENTERITIS, an inflammation of the intestines. Consult a physician if symptoms are severe or if blood, pus or mucus is noted in the stool.

● **Nausea and vomiting that occur suddenly and are associated with crampy abdominal pain and diarrhea** may suggest FOOD POISONING, bacterial contamination of food. Symptoms generally affect most or all of the persons who have eaten the same food. The symptoms usually last for a day or so. Rest and drink extra fluids, but consult a physician if symptoms persist. If such symptoms are swiftly followed by difficulty seeing, speaking or swallowing, or by sudden weakness or paralysis, the cause may be BOTULISM, food poisoning caused by *Clostridium* bacteria. Consult a physician immediately if you suspect this ailment.

● **Nausea and vomiting that are accompanied by persistent or severe abdominal pain and tenderness** can indicate any of several serious digestive disorders.

If nausea and vomiting occur together with pain in the upper right side of the abdomen, in the back or in the right shoulder, the cause may be BILIARY COLIC, partial blockage of the bile duct by a gallstone. Pain and nausea usually occur suddenly, persist for several hours and then subside. If the pain passes, you can wait until the next day to see the doctor; but if it worsens or if the yellow skin of jaundice develops, you should go immediately.

If nausea and vomiting occur together with pain in the upper right side of the abdomen and with fever, the cause may be ACUTE CHOLECYSTITIS, an inflamed gall bladder. The pain may feel as if it is boring through to the back. If such symptoms occur with the yellow skin of jaundice, the cause may be ACUTE CHOLANGITIS, a complete blockage of the bile duct by gallstones. Consult a physician within a few hours if you suspect either of these disorders.

If nausea and vomiting occur together with pain in the lower right side of the abdomen and with fever, appetite loss and constipation, the cause may be ACUTE APPENDICITIS, an infected appendix. Pain usually begins in the middle of the abdomen and later moves to the lower right. Consult a physician immediately.

If nausea and vomiting occur together with severe midabdominal pain that bores through toward the back, the cause may be ACUTE PANCREATITIS, an inflamed pancreas. Other symptoms include sweating, abdominal swelling, and incapacitating weakness. This condition is most often due to GALLSTONES or alcoholism. Consult a physician immediately.

If nausea and vomiting occur with severe midabdominal pain and abdominal swelling, they may suggest ACUTE INTESTINAL OB-STRUCTION, a blockage of the bowel with back-up of fluid and gas. Consult a physician immediately.

● **Nausea and vomiting accompanied by fever, appetite loss, changes in taste or smell, fatigue and the yellowed skin of jaundice** may indicate HEPATITIS, an infection that inflames the liver. Additional symptoms include headache, cough and aching muscles or joints. Jaundice may occur together with or after the other symptoms. Consult a physician at the first sign of jaundice.

● **Nausea and vomiting that are accompanied by throbbing head pain on one side of the head** may indicate COMMON MIGRAINE, painfully dilated blood vessels, also called SICK HEADACHE. There may also be sensitivity to light and sound, or abdominal pain. If such symptoms are preceded by a 15- to 30-minute period during which unusual sights or feelings are experienced, the disorder may be the less common CLASSIC MIGRAINE. Consult a physician if symptoms are severe or recurrent.

● **Nausea and vomiting that are accompanied by severe, gripping or crushing chest pain,** shortness of breath and sweating may indicate MYOCARDIAL INFARCTION (HEART ATTACK), a blockage of blood flow to the heart (see CHEST PAIN). Get medical attention as soon as possible.

RECTAL BLEEDING. *See STOOL CHANGES*

REGURGITATION. The sudden appearance of partially digested food in the back of the throat or mouth is an unpleasant symptom that frequently occurs without nausea or vomiting. This is usually caused by overeating and is nothing to worry about unless it is a daily problem or blood is present in the regurgitated food.

● **Painless regurgitation that occurs frequently in an elderly person** may indicate ZENKER'S DIVERTICULUM, the formation of a tiny pouch in the upper part of the tube that carries food from the mouth to the stomach. Consult a physician within a few days.

● **Regurgitation that occurs with a warm or burning pain behind the breastbone** may suggest HIATAL HERNIA, a protrusion of the stomach through a hole in the diaphragm. Or it may indicate REFLUX ESOPHAGITIS, an irritation of the lining of the food tube by stomach juices. Symptoms may be aggravated by eating heavy meals, lying down or straining. Eat less and elevate the head of your bed about six to eight inches. Try to lose weight, for this will lessen pressure on the stomach.

● **Regurgitation that is associated with either sharp or dull pain behind the breastbone, or with blood in the regurgitated food,** may suggest ACHALASIA, a dilated and ineffective food tube; DIFFUSE ESOPHAGEAL SPASM, painful muscular contraction of the

food tube; INFLAMMATORY STRICTURE, a scarred and narrowed food tube; or ESOPHAGEAL CANCER, a tumor in the food tube. Consult a physician immediately if pain is severe or if blood is present in the regurgitated food.

STOOL CHANGES. Most stool changes are temporary and need cause no concern. But those changes that persist for more than three days—especially blood in or on the stool—should send you to the doctor.

● **Blood that appears as a bright red staining of the toilet tissue or stool** usually indicates a minor disorder involving the sensitive anal tissue.

If bloody stools are associated with painful, uncomfortable or incomplete bowel movements, or with an itching or burning sensation around the anus, the cause may be HEMORRHOIDS, swollen rectal or anal veins. A firm lump protruding through the anus suggests a THROMBOSED HEMORRHOID, a blood clot in a swollen rectal or anal vein. Both conditions normally respond to home treatment (see ANAL DISCOMFORT). However, any rectal bleeding should be called to the attention of a doctor, for it may be a symptom of a serious ailment.

● **Blood-tinged or grossly bloody diarrhea** may indicate one of several serious bowel disorders.

If bloody diarrhea occurs with pus and mucus admixed, the problem may be ULCERATIVE COLITIS, an inflammatory bowel disease that can also cause abdominal pain, fever and weight loss. Consult a physician within several hours.

If bloody diarrhea occurs with fever and crampy abdominal pain, the cause may be SHIGELLOSIS or DYSENTERY—two varieties of GASTROENTERITIS, an inflammation of the intestines. Consult a physician within several hours.

If bloody diarrhea is associated with severe abdominal pain in an elderly person, the cause may be ISCHEMIC COLITIS, insufficient blood flow to the intestine. Consult a physician immediately.

● **Blood that makes the stool maroon or a tarry black** often indicates bleeding high in the digestive tract and requires professional attention. However, black stools can also be caused by eating licorice or by taking iron pills or medicines containing bismuth.

If loose, maroon stools occur, the cause may be DIVERTICULITIS, bleeding from a small pouch that has formed in the large intestine. Consult a physician immediately.

If black, tarry stools occur, they may indicate bleeding from an ULCER, a hole in the lining of the stomach or upper intestine; or from GASTRITIS, an inflamed stomach lining. Consult a physician immediately.

If firm black or unusually thin stools are associated with constipation or a change in bowel habits, the cause may be COLONIC

CANCER, a tumor in the large intestine. Consult a physician within a day or so.

● **Stools that are unusually loose, bulky, foul-smelling, greasy or pale** may suggest LACTOSE INTOLERANCE, the inability of the intestine to digest dairy products; or GLUTEN INTOLERANCE, the inability to digest proteins in wheat, rye, barley and oats. Additional symptoms include weight loss, flatulence and abdominal swelling. Try to avoid suspected foods *(pages 64-65),* but consult a physician within several days.

SWALLOWING DIFFICULTY. This symptom—known medically as dysphagia—is sometimes described as food sticking in the chest. Difficulty swallowing may occur suddenly and painfully, but it can also develop gradually, without pain, over several weeks or months. Temporary swallowing difficulties may be caused by nothing more serious than the sore throat of a common cold or influenza: Gargle with warm salt water, take aspirin and consult a physician if pain is severe or if fever or swollen glands in the neck are present. Occasionally, however, difficulty swallowing can signal a condition that demands prompt medical attention.

● **Difficulty swallowing that occurs with chest pain** can signal several disorders, rarely indicating a problem with the heart.

If difficulty swallowing is accompanied by chest pain, coughing or regurgitation of food into the mouth, the cause may be a disorder of the esophagus, the tube that carries food from the mouth to the stomach. These disorders include ACHALASIA, a dilated and ineffective food tube; a DIFFUSE ESOPHAGEAL SPASM, a painful muscular contraction of the food tube; an INFLAMMATORY STRICTURE, a scarred and narrowed food tube; or ESOPHAGEAL CANCER, a tumor in the food tube. Consult a physician immediately if symptoms are severe, but you can wait until the next day if they are not.

● **Difficulty swallowing that is associated with recurrent coughing episodes** may indicate ASPIRATION, food passing into the airway and lungs. This condition can lead to pneumonia. Consult a physician if coughing occurs after each swallow.

● **Difficulty swallowing that is associated with coughing or the painless regurgitation of food into the mouth** may suggest ZENKER'S DIVERTICULUM, a pouch that sometimes forms in the upper part of the esophagus. The condition primarily affects the elderly. Consult a physician within a day or so.

VOMITING. *See NAUSEA AND VOMITING*

VOMITING BLOOD. Called hematemesis by doctors, this problem is a serious symptom that always demands immediate medical attention. The blood is usually bright red, but if it has been altered by stomach juices it can be dark brown or resemble coffee grounds. Vomiting blood most often occurs without warning, but may be preceded by abdominal pain or the appearance of black, tarry stools.

If vomiting blood is accompanied by pain in the pit of the stomach, between the navel and ribs, the problem may be GASTRITIS, an inflamed stomach lining. The inflammation is most often the result of excessive alcohol consumption, infection or sensitivity to aspirin or other drugs.

If vomiting blood occurs in a person who has experienced recurrent abdominal pain that feels like a gnawing or burning ache in a small area, the cause may be an ULCER, a sore in the lining of the stomach or upper intestine.

If vomiting blood occurs during or just after an episode of prolonged retching or nonbloody vomiting, it may be a sign of MALLORY-WEISS SYNDROME, also known as an esophageal laceration, a tear in the lining of the food tube.

WEIGHT CHANGES. Changes in weight usually are easily explained—stemming from overindulgence, a diet or a minor illness such as influenza. Unexplained weight change, however, can be an indication of a serious problem.

● **Weight gain that is not readily explained** may suggest the inadequate heart output of CONGESTIVE HEART FAILURE; CIRRHOSIS, a scarred liver that often results from alcoholism; or NEPHROTIC SYNDROME, a kidney disease that can cause severe protein loss and, eventually, malnutrition. In such cases, the weight gain usually is accompanied by a visible swelling of the face, legs or abdomen or, in CIRRHOSIS, by jaundice (yellow skin). Consult a physician.

● **Weight loss that occurs in spite of increased food intake** may indicate HYPERTHYROIDISM, an overactive thyroid gland; the condition is normally accompanied by nervousness or a persistent warm feeling. Such weight loss might also indicate DIABETES, improper sugar processing that often causes frequent urination and unusual thirst. Consult a physician within a few days.

If weight loss and increased food intake are associated with chronic diarrhea, or with bulky, foul-smelling, or greasy, yellow stools, the cause may be MALABSORPTION, improper absorption of food from the intestines. Consult a physician within a day or two.

● **Weight loss that is accompanied by persistent appetite loss** may indicate CANCER; or it may arise from UREMIA, kidney failure that may also cause nausea, vomiting and behavior changes such as sleep or memory disturbances. Consult a physician within a day.

If weight loss is associated with absence of appetite or severe dieting in an adolescent, particularly a girl, it may suggest ANOREXIA NERVOSA, a psychological problem that can cause life-threatening malnutrition. Consult a physician within a day.

Bibliography

BOOKS

Alvarez, Walter C., *Nervousness, Indigestion and Pain*. Harper & Brothers, 1954.

Berland, Theodore, et al., *Living with Your Colitis and Hemorrhoids and Related Disorders*. St. Martin's, 1975.

Berland, Theodore, and Mitchell A. Spellberg, *Living with Your Ulcer*. St. Martin's, 1971.

Crile, George, Jr., *Surgery: Your Choices, Your Alternatives*. Delacorte, 1978.

Crohn, Burrill B., *Understand Your Ulcer: A Manual for the Ulcer Patient*. Sheridan House, 1950.

Denney, Myron K., *Second Opinion*. Grosset & Dunlap, 1979.

The Editors of Consumer Reports Books, *The Medicine Show*. Pantheon Books, 1980.

Eiseman, Ben, *What Are My Chances?* Saunders, 1980.

Eisenberg, M. Michael, *Ulcers*. Random House, 1978.

Galton, Lawrence:
The Patient's Guide to Surgery: How to Make the Best of Your Operation. Hearst Books, 1976.
Save Your Stomach. Crown, 1977.

Grossman, Morton I., ed., *Peptic Ulcer: A Guide for the Practicing Physician*. Year Book Medical Publishers, 1981.

Holt, Robert Lawrence, *Hemorrhoids: The Problem, Personal Treatment, Medical Treatment*. California Health Publications, 1977.

Lehmann, Helmut T., ed., *Luther's Works, Volume 48, Letters I*. Transl. by Gottfried G. Krodel, Fortress, 1963.

Long, James W., *The Essential Guide to Prescription Drugs: What You Need to Know for Safe Drug Use*. Harper & Row, 1980.

Melluzzo, Paul J., and Eleanor Nealon, *Living with Surgery: Before and After*. Lorenz, 1979.

Nolen, William A.:
The Making of a Surgeon. Random House, 1968.
A Surgeon's World. Random House, 1970.

Nugent, Nancy, *How to Get Along with Your Stomach*. Little, Brown, 1978.

Serino, G. S., *Your Ulcer: Prevention, Control, Cure*. Lippincott, 1966.

Sleisenger, Marvin H., and John S. Fordtran, *Gastrointestinal Disease: Pathophysiology, Diagnosis, Management*. W. B. Saunders, 1978.

Spiro, Howard M., *Clinical Gastroenterology*. Macmillan, 1977.

Wolf, Stewart, *The Stomach*. Oxford University Press, 1965.

PERIODICALS

Brady, Joseph V., "Ulcers in 'Executive' Monkeys." *Scientific American*, October 1958.

Burakoff, Robert, "An Updated Look at Diverticular Disease." *Geriatrics*, Vol. 36, No. 3, March 1981.

Cave, D. R., et al., "Evidence of an Agent Transmissible from Ulcerative Colitis Tissue." *The Lancet*, June 19, 1976.

Drossman, Douglas A., et al., "The Irritable Bowel Syndrome." *Gastroenterology*, Vol. 73, No. 4, 1977.

"The Epic of Hepatitis B: Victim's Blood Yields Landmark Vaccine." *Medical World News*, October 26, 1981.

Epstein, Max, "Endoscopy: Developments in Optical Instrumentation." *Science*, Vol. 210, October 17, 1980.

Fanning, J. C., et al., "Converting a Stomach to a Uterus: The Microscopic Structure of the Stomach of the Gastric Brooding Frog Rheobatrachus silus." *Gastroenterology*, Vol. 82, No. 1, January 1982.

Farmer, Richard G., et al., "Studies of Family History Among Patients with IBD." *Clinics in Gastroenterology*, Vol. 9, No. 2, May 1980.

"For Answers to Diabetes, Gallstones, Arthritis, Ask the Pima Indians." *Medical World News*, March 3, 1980.

"Getting to the Seat of the Problem." *FDA Consumer*, September 1980.

Goldstein, Franz, "Inflammatory Bowel Disease—Better Prospects." *Consultant*, December 1980.

Grossman, Martin B., "Gastrointestinal Endoscopy." *Clinical Symposia*, Vol. 32, No. 3, 1980.

Janowitz, Henry D., "Crohn's Disease—50 Years Later." *The New England Journal of Medicine*, Vol. 304, No. 26, June 25, 1981.

Kolff, Willem J., "First Clinical Experience with the Artificial Kidney." *Annals of Internal Medicine*, Vol. 62, No. 3, March 1965.

Mendeloff, Albert I., "The Epidemiology of Inflammatory Bowel Disease." *Clinics in Gastroenterology*, Vol. 9, No. 3, May 1980.

Mills, John, "Acute Gastroenteritis: The When, What, and How of Treatment." *Modern Medicine*, Vol. 47, No. 18, October 30-November 15, 1979.

Moog, Florence, "The Lining of the Small Intestine." *Scientific American*, November 1981.

Painter, Neil S., and Denis P. Burkitt, "Diverticular Disease of the Colon: A Deficiency Disease of Western Civilization." *British Medical Journal*, May 22, 1971.

Pendower, J. E. H., "Spontaneous Disappearance of Gall-stones." *British Medical Journal*, 1964.

Perrillo, Robert P., and Richard D. Aach, "Acute Viral Hepatitis: Current Concepts and Practical Guidelines for the Primary Physician." *Practical Gastroenterology*, Vol. 5, No. 2, March/April 1981.

Pietrusko, Robert G., "Use and Abuse of Laxatives." *American Journal of Hospital Pharmacy*, March 1977.

Rutkow, Ira M., "Surgical Rates in the United States: 1966 to 1978." *Surgery*, Vol. 89, No. 2, February 1981.

Sachar, David B., et al., "Aetiological Theories of Inflammatory Bowel Disease." *Clinics in Gastroenterology*, Vol. 9, No. 2, May 1980.

Sampliner, Richard E., et al., "Gallbladder Disease in Pima Indians: Demonstration of High Prevalence and Early Onset by Cholecystography." *The New England Journal of Medicine*, Vol. 283, No. 25, December 17, 1970.

Schoenfield, Leslie J., "Diseases of the Gallbladder and Biliary System." *Practical Gastroenterology*, Vol. 5, No. 2, March/April 1981.

Silverstein, Fred, et al., "Endoscopic Hemostasis Using Laser Photocoagulation and Electrocoagulation." *Digestive Diseases and Sciences*, Vol. 26, No. 7, July 1981.

Standards Committee of the American Society of Transplant Surgeons, "Current Results and Expectations of Renal Transplantation." *Journal of the American Medical Association*, Vol. 246, No. 12, September 18, 1981.

Sturdevant, Richard A. L., "Increased Prevalence of Cholelithiasis in Men Ingesting a Serum-Cholesterol-Lowering Diet." *The New England Journal of Medicine*, Vol. 288, No. 1, January 4, 1973.

Szmuness, Wolf, et al., "Hepatitis B Vaccine: Demonstration of Efficacy in a Controlled Clinical Trial in a High-Risk Population in the United States." *The New England Journal of Medicine*, Vol. 303, No. 15, October 9, 1980.

Wenckert, Anders, and Bertil Robertson, "The Natural Course of Gallstone Disease: Eleven-Year Review of 781 Nonoperated Cases." *Gastroenterology*, Vol. 50, No. 3, March 1966.

Picture credits

The sources for the illustrations that appear in this book are listed below. Credits for the illustrations from left to right are separated by semicolons, from top to bottom by dashes.

Cover: Photo by Fil Hunter, Art by Trudy Nicholson. 7: Al Freni, courtesy Gordon Mestler. 9: From *Tissues and Organs: A Text-Atlas of Scanning Electron Microscopy*. By Richard G. Kessel and Randy H. Kardon. W. H. Freeman and Company. © 1979. 10: Culver Pictures. 11: Culver Pictures—Sovfoto. 12: Circus World Museum, Baraboo, Wisconsin. 13: Henry Grossman for *People*. 14: Southern Illinois University, School of Medicine. 17-20: Drawings by Trudy Nicholson. 24: The British Museum, London. 26-33: © 1974 Lennart Nilsson from *Behold Man*, Little, Brown and Co. Drawings by Trudy Nicholson. 35: Leonard Barbiero. 36: Roland Michaud-Rapho, Paris. 38: Drawing by John Drummond. 40: *Royal Society of Tropical Medicine and Hygiene* 74:429-433, 1980, courtesy Robert L. Owen, M.D., V.A. Medical Center, San Francisco. 42, 43: John Neubauer. 45: National Library of Medicine. 46: Children's Hospital of Pittsburgh and Georgetown University Hospital-National Capital Poison Center; © Rocky Mountain Poison Foundation, Intermountain Regional Poison Control Center, University of Utah. 49: Library of Congress. 59: John Neubauer. 60: Sturgis Grant Productions, Inc., courtesy Smith Kline & French Laboratories. 61: Philadelphia Museum of Art: Given anonymously. 62: Drawings by John Drummond. 64, 65: Drawings by Trudy Nicholson. 73: © Gary Gladstone from The Image Bank. 77: Juan Lechago, M.D., Ph.D. 81: Walter Reed Army Institute of Research. 83: © M. J. Tyler, Zoology Department, University of Adelaide, North Terrace, Adelaide, South Australia, Australia. 84: David Fleischer, M.D. 86: John Senzer, courtesy The University of Chicago Medical Center. 87: John Senzer, courtesy The University of Chicago Medical Center; A. S. Cigtay, M.D. 88, 89: Drawings by John Drummond. 92-94: John Senzer, courtesy The University of Chicago Medical Center. 95: B. H. Gerald Rogers, M.D.; Walter Hilmers Jr. from H J Commercial Art—John Senzer, courtesy The University of Chicago Medical Center. 96-100: John Senzer, courtesy The University of Chicago Medical Center. Drawings by Walter Hilmers Jr. from H J Commercial Art. 101: Scintiphotos, courtesy Terry F. Brown, Section of Nuclear Medicine, The University of Chicago Medical Center,
drawings by Walter Hilmers Jr. from H J Commercial Art—John Senzer, courtesy The University of Chicago Medical Center. 102, 103: John Senzer, courtesy The University of Chicago Medical Center (2); Walter Hilmers Jr. from H J Commercial Art; John Senzer, courtesy The University of Chicago Medical Center— W. E. Jensen, Ph.D. 104, 105: John Senzer, courtesy The University of Chicago Medical Center. 107: Olympus Corporation of America. 108: Merck Sharp & Dohme. 110: Yukihiko Nose, M.D., Ph.D., Department of Artificial Organs, Cleveland Clinic. 114: Dan McCoy from Rainbow. 115: The University of Utah, Dialysis in Wonderland. 117: John Senzer, courtesy The University of Chicago Medical Center. 118: Wide World, courtesy Åke Håkanson, Reportagebild. 120: John Senzer, courtesy The University of Chicago Medical Center. 122-129: Linda Bartlett. 131: © 1981 Howard Sochurek. 133: Elizabeth Bennion, courtesy of St. John's College, Oxford, England. 136: Drawing by John Drummond. 140: From the collection of Gordon E. Mestler, Downstate Medical Center, Brooklyn, New York. 142: Drawing by Trudy Nicholson. 144: Bruce Griffeth, Atlanta Regional Organ Procurement Agency. 146-163: Richard Anderson, courtesy Johns Hopkins Hospital, Baltimore.

Acknowledgments

The index for this book was prepared by Barbara L. Klein. For their help in the preparation of this volume, the editors wish to thank the following: Linda Andrus, Olympus Corporation of America, New Hyde Park, N.Y.; Nirmala Auerbach, Yale-New Haven Hospital, Conn.; Dr. Wilma Bias, Johns Hopkins Hospital, Baltimore, Md.; Terry F. Brown, University of Chicago Medical Center; Dr. James Burdick, Johns Hopkins Hospital, Baltimore, Md.; Deborah Diggs, Johns Hopkins Hospital, Baltimore, Md.; Anne Eggleton, University of Chicago Medical Center; Dr. Barbara Fivush, Johns Hopkins Hospital, Baltimore, Md.; Dr. David Fleischer, Veterans Administration Hospital, Washington, D.C.; Dr. Paul Frank, University of Chicago Medical Center; Dennis Giesing, Marion Laboratories, Kansas City, Mo.; Dr. Michael S. Gold, Washington Hospital Center, Washington, D.C.; Dr. Mark Greenberg, University of Chicago Medical Center; Ann Grinham, Marion Laboratories, Kansas City, Mo.; Jane A. Gross, American
Digestive Disease Society, New York City; Martin I. Hassner, American Digestive Disease Society, New York City; Dr. David Hutchin, Johns Hopkins Hospital, Baltimore, Md.; Dr. William Hutchins, University of Chicago Medical Center; Dr. Dick Johannes, National Institutes of Health, Bethesda, Md.; Wilson Kierstead, New York City; Dr. Joseph Kirsner, University of Chicago Medical Center; Lee and Marshall Koonin, Lifeline Foundation, Sharon, Mass.; Dr. James E. Lander, Richmond, Va.; Dr. Juan Lechago, UCLA School of Medicine, Torrance; Dr. Bernard Levin, University of Chicago Medical Center; Dr. Michael Levitt, Veterans Administration Medical Center, Minneapolis, Minn.; Deborah MacFarlane, University of Chicago Medical Center; Dr. Fray Marshall, Johns Hopkins Hospital, Baltimore, Md.; Dr. Robert McLean, Johns Hopkins Hospital, Baltimore, Md.; Miriam and John McMaster, North East, Md.; Marion White McPherson, University of Akron, Ohio; G. E. Mestler, Downstate Medical Center, Brooklyn, N.Y.; Dr. Wesley Norman, Georgetown University, Washington, D.C.; Dr. Yukihiko Nose, Cleveland Clinic, Ohio; Gene A. Pierce, Southeastern Organ Procurement Foundation, Richmond, Va.; John A. Popplestone, University of Akron, Ohio; Ibrahim Pourhadi, Library of Congress, Washington, D.C.; Miriam Ratner, American Digestive Disease Society, Bethesda, Md.; Dr. Judy Reid, Marion, Va.; Dr. B. H. Gerald Rogers, University of Chicago Medical Center; Dr. David B. Sachar, Mount Sinai School of Medicine, New York City; Dr. Philip Schmidt, Brooklyn, N.Y.; Dr. Arthur Sicular, Mount Sinai Medical Center, New York City; Rose Ann Soloway, Georgetown University Hospital, Washington, D.C.; Penelope Steiner, National Foundation for Ileitis and Colitis, Inc., New York City; Dr. John H. Texter, Jr., Southern Illinois University School of Medicine, Springfield; William Troutman, American Association of Poison Control Centers, Albuquerque, N. Mex.; Dr. Harold J. Tucker, Baltimore City Hospital, Md.; Alan Wachter, SmithKline Corporation, Philadelphia, Pa.; John Warner, University of Utah, Salt Lake City; Dr. H. David Watts, University of California School of Medicine, San Francisco; C. Philip Wilson, New York City. Special diets were adapted from *Nutrition Care Manual* prepared jointly by Yale-New Haven Medical Center, the Hospital of St. Raphael, West Haven Veterans Administrative Medical Center and the Waterbury Hospital Health Center.

Index